West Bromwich Corporation

Derby City

BUS

ic pride.
run vehicles of the past.

Chesterfield Corporation

Walsall Corporation

MIDBUS

SOME ASPECTS OF MIDLAND BUS AND
COACH OPERATIONS

David Gladwin

K.A.F. BREWIN BOOKS

First published in December 1992 by
K.A.F. Brewin Books, Studley, Warwickshire.

**Birmingham Corporation
Tramway and Omnibus Department**

NOTICE TO CONDUCTORS

The use of the words "Ma", "Ta" and similar expressions by Conductors has become very prevalent. Whilst Conductors may not consider this objectionable it is deemed an impertinence by passengers, and the use of these terms must therefore cease at once.

*By Order,
A.C. Baker.*
General Manager.

Head Office
November 1935

© David Gladwin 1992

All rights reserved

ISBN 0 947731 91 1

Designed and Typeset by Supaprint (Redditch) Ltd
and made and printed in Great Britain.

ACKNOWLEDGEMENTS

Acknowledgements in a book of this type cannot do justice to the dozens of men and women who provided some part of the jigsaw. Ross Hamilton, Reg Ludgate, A.P. Newland, Ned Williams, K. Long of the Burton Daimler Group are a few of the individuals. Members of the Omnibus Society, the PSV Circle, the H.C.V.S. and British Bus Preservation Group and many other transport societies joined with various Museums including the Oxford Bus Preservation Society at Handborough, Oxon, the Midland Bus Museum at Wythall, Worcs., and the London Transport Museum, in adding facets of knowledge. Libraries that particularly went out of their way included Birmingham Reference, Huddersfield Central and Newcastle City. The Birmingham Transport Historical Group kindly agreed to our using material from Coventry Transport Vol. II by Denton & Groves. The Editors of Bus Fayre, Coach & Bus Week, Coaching Journal & Bus Review, Commercial Motor, Vintage Roadscene and Vintage Commercial Vehicle Magazine have all given permission to reproduce copyright material. Particular thanks must go to Roger Kidner and the Oakwood Press for writing and publishing very many thoroughly useful transport titles and allowing us to use extracts. I also thank John Heath and the Paddock Press for permission to reproduce items from their publications. Mrs. Mona Morgan wrote, and The Gomer Press published, "Growing Up in Kilvert Country" and without this volume our knowledge of earlys PSVs would be much poorer.

Photographically we owe thanks for permission to reproduce copyright material to M. Allan - 81 (lower); W.J. Haynes - 41; J. Heath - 42-44, 88; A. Jarosz - 51 (upper), 69 (lower) 76 (top); D.A. Jones - 39; K. Jenkinson - 69 (upper); R.W. Kidner - 8 (lower), 52 (upper), 53 (upper); C. Meir - 65; A. Simpson - 38, 39; J. Wilson Front Cover, 54-55, 80; J-J Wright - 15 (lower). R.H.G. Simpson has available an amazing choice and not unsurprisingly has provided a dozen or so photographs scattered throughout the book. The clever drawings on pages 89 and 90 are by Steve Hallam of Farnborough who has a keen eye for the foibles of bus and coach drivers.

Finally, my wife Joyce has an unenviable problem living with a transport employee-cum-writer. She has all the dreary jobs like typing and little of the credit she deserves.

The maximum speed permitted by Public Service Vehicles remained 12 m.p.h. (19.3 k.p.h.) until 1 October 1928. After then vehicles fitted with pneumatic tyres on both axles were allowed to travel at 20 m.p.h. (32.2 k.p.h.)

From 1 January 1931 this speed was increased to 30 m.p.h. (48.3 k.p.h.). This remained even in 1959 when the M1 was opened from London to Birmingham. Paradoxially, no speed limit whatsoever applied on motorways. From May 1961 P.S.Vs were permitted to travel at 40 m.p.h. (64.4 k.p.h.) on all roads except motorways which remained without any limit. In August 1966 the general limit increased to 50 m.p.h. (80.5 k.p.h.) but from the following year 70 m.p.h. (112.6 k.p.h.) limits applied to all vehicles on motorways.

Front Cover:
Modern vehicle, cricketing weather, traditional manufacturer in a historical town. Wood Street, Stratford-upon-Avon, August 1991. Leyland Olympian, number 963 within the Midland Red South fleet has been most attractively painted in classical 'Stratford Blue' colours. Chassis classification ONLXB/1RH, bodywork by Eastern Coachworks with surprisingly comfortable dual purpose seating; new 1986.

John Wilson

Back Cover:
Once, as a front line coach, number 789 but latterly number 7 in the Midland Red South fleet, BVP 789V, a 1980 Leyland Leopard PSU3E/4R with Willowbrook 49 seat bodywork, passed to Vanguard Travel in July 1989. Behind are a pair of ex-West Midlands P.T.E. (now West Midlands Travel) Mk.II DR 102/27 73 seater Metrobuses. Photographed at Fairfax Street, Coventry, at about midday on 1st February 1992 during a spell of wet and foggy weather. Vanguard are an associate company of Midland Red South within the Western Travel Group.

Frontispiece

Midlands' transport personified at The Midlands' Museum of Transport, Wythall, in 1992. HOV 685, ex-Birmingham City Transport is a 1948 Leyland Titan PD2 with Brush body-work. The single decker, RC 4615, ex-Trent Motor Traction, has a 1937 AEC Regal chassis, with a 1950 Willowbrook body. KON 311P, from the West Midlands P.T.E. fleet is a 1976 Daimler Fleetline FE30 ALR with Metro-Cammell built bodywork.

Greetings to all our Distributors, Customers and Friends at home and overseas. With best wishes for Health, Happiness and Prosperity in 1950

"I have tremendous faith in the inherent commonsense of the average Britisher and I hope ere long to see this great country of ours restored by industry from a 'fool's paradise' to the position she held among the nations of the world, built up by industry in the age when little was taxed, when the £ was worth 20s; when it was recognised that 'an ounce of practice was worth a ton of theory'; when to work hard was a credit and to slack a crime; when an Englishman's word was his bond; and one heard the expression 'As safe as the Bank of England'. Otherwise we shall slide from a Welfare State to a "Farewell State".

Mr. Sydney S. Guy, chairman and managing director, Guy Motors Ltd., in a statement to shareholders, 7 January 1950.

INTRODUCTION

This book is about the Midlands but the reality is different, for the Midlands cannot be considered in isolation. All the manufacturers, whether Sentinel in Shrewsbury or Brush in Loughborough, had to export or die, even if those exports were only to London or Edinburgh or Tavistock.

Similarly, particularly before the era of the repmobile and the motorway, public transport carried people about their business. They might take the train from London to Wolverhampton but when they reached the High or Low Level Stations, they took the bus. Workmen never expected to drive to 'The Austin' or to Morris's or Standards but instead took either one of a Corporation's vehicles or in such far flung places as Tamworth or Bromsgrove made use of, and helped to keep solvent, a private operators "Works' Service".

Holidays were not taken in Costa del Noisy or Zanzibar but instead in Blackpool, Yarmouth or Weymouth reached, more often than not, by a coach from the local operator's yard whose licence restriction often led to the notation "Saturdays only, July and August".

But another aspect of coach travel was the desire of people to travel from say, London to Liverpool via, perhaps, Birmingham or Stratford or some other Midland town. Until the late 1930s and in some cases far later in the century it was quite an accepted practice to split such a long journey into two or three days travel, overnighting at some hotel which appealed or (more economic and socially correct) at Aunt Phoebe's or Uncle Jack's. This habit lingered on until piston engined aircraft days when a passenger would travel from London to New York and happily spend a night at Paris or Frankfurt and again at Boston. The cost was little more than the direct fare.

Insofar as the photographs are concerned either the vehicles chassis was built in the Midlands, whether Guy, Daimler or Maudslay, or the bodywork came from a Midlands' firm, among them Buckingham, Brush, Metalcraft; or the company itself was Midland based. One could in one vehicle get pure Black Country engineering - Guy chassis, Meadows engine, Guy body; hybrids - Crossley with Mulliner (Northampton) body or Sunbeam/Park Royal. There were even, albeit misguided, operators who, Midlanders to their very boots, bought all—Leyland PD2s or AEC/Roe.

For political reasons Corporations had a rather restricted choice of vehicle and it was a wise engineer who found out the prejudices of his finance committee before he put forward silly ideas. For example, in times of severe recession whatever the merits of AEC chassis, Coventry's 'Main Man', who might be God in the garage, would prove to be very mortal if he persisted in trying to buy AEC rather than Daimler chassis. Local employment = rates and council house rents and shopkeepers ringing tills and a certain pride in their town, all of which weighed very heavily in the balance. Unemployment meant Poor Law Relief, and hospital beds closing, and education slipping as children were kept away for lack of boots, and streets not being cleaned, and vagrancy. This was the scenario only 40 years ago. In a town like Wolverhampton the transport committee had an appalling problem for they had to decide from whom to buy, say, ten vehicles in one year. Should they support Guy? Or Sunbeam? Or Star? Do they buy two or three of each and strangle for the lack of spares or put all their eggs in one basket, risking the other companies going to the wall? It was a very real problem, as it would be noted that Guy had had an order and the engineer in Nottingham or Northampton might well wonder what was wrong with Sunbeam and mentally cross them off the list.

Birmingham bought fifty of 83 Morris Commercial Imperial chassis built and they were, by all accounts, mobile disasters. But they (and a handful of Dictator chassis) were bought in the dreadful years of 1931-4 when, no doubt, these orders kept the Adderley Park Works in work; albeit even so, three day weeks became necessary. But three days pay to the men was better than none. Incidentally, bodies for BCT vehicles in that period were an odd mixture, although Metro-Cammell and the Birmingham Carriage & Wagon Company got the bulk of the work.

The sub-title of the book may also explain some apparently anomalous features. The work of the Traffic Commissioners is not always apparent but in this day of deregulation without their activities mobile death-traps would be all too commonplace. In the 1950s before the MOT test, most impecunious companies had an old, often pre-war, hulk lying around which was only rejuvenated in the peak summer season - at 30 mph (50 kph) the power of tatty drum brakes to stop 6 or 7 tons was not too vital; but at today's speeds would be downright lethal.

Maintenance is very relevant for more and more quite intelligent people are turning their minds and bodies to the problems of restoring elderly coaches and buses, so a look at how work was carried out by bodyworks in the days when their vehicles were new seems desirable. The bus and coach industry has always had a human face and nowhere was this more apparent than in wartime. Bravery was almost commonplace but what of the church who protested at the loss of revenue from their congregation when in the height of the U-boat campaign a Corporation were forced to reduce bus services due to the lack of fuel?

So, buses, coaches, and those dreams of vehicles, trolleybuses, all have their place but also encapsulated in text and photographs are people, the people who built them, who supplied components, who operated them, who drove them indeed, and, for without the one the other cannot exist, the passengers who rode in, and in early days on, public service vehicles. There is the steam omnibus with the crew looking more like railwaymen, and there is the luxury coach with uniformed driver looking more like a chauffeur and there is, insofar as it is possible, everything in between. Obviously we accept there are omissions and the publisher and author alike would invite comments or ideas for possible future editions.

CONTENTS

	Page
Quadrille - Classic Coach Design	2
Midscene - Link Machines	4
Mobile Catering 1920s style	9
Thumbnail Sketch 1 - Evertons of Droitwich	10
London to Cardiff	11
London and Newcastle	14
Tiger, Tiger, Burning Bright	16
Via the Heart of England	18
Thumbnail Sketch 2 - RHMS Coventry	20
The Role of the Traffic Commissioner by J. Mervyn Pugh	21
Birmingham to Belfast	26
NorWest goes East	30
The Rural Problem	31
The Market Bus	34
Who's a Little Sunbeam then by Andy Simpson	36
Six Towns' Transport	41
Derby & District Bus Days	42
Brum's Bygone Buses	48
By Easy Stages by John Reohorn	50
Midland Red in the Country	51
Midscene - Stratford-upon-Avon	54
Transport at War by Judy-Joan Wright	56
Potteries Preservation	65
Thumbnail Sketch 3 - Brown's Blues, Markfield	66
From West Midlands to Everywhere	67
Omnibus & Coach 1929	71
Midscene - Roadside Cafes	76
Everyone a Little Gem	77
Golden Days of Advertising	82
Thumbnail Sketch 4 - Burton-upon-Trent Corporation	85
Top to Bottom and Bottom to Top	86
Midscene - People	88
Humour	89
Tickets from a Conductor's Eye by Conductor 0129	91
John Buckingham, Brum's Bus Builder	96
Midlands Made . . . or Partially so	98
Walsall Double Deckers	100
In memory of Daimlers	102
Appendices	104
Index	106

QUADRILLE - CLASSIC COACH DESIGN

This coach exudes the late 1930s in its styling and the body design from the thick pillar behind the cab is pure Duple c. 1937. The front has all the appearance of a different vehicle and here lies the clue. When James Whitson of West Drayton commenced their coach building activities what more logical than there would be a drift of staff from Duples down the road at Hendon! This, then, was a 1947 offering from Whitsons on a Maudslay Marathon Mark III chassis. Seats are a so-called dual-purpose pattern, low backed and almost bus-like, radio is obviously fitted, but there is an indefinable lack of styling and trim for a coach of the period. Later Whitsons were to show flashes of originality but to little avail as orders dried up in the 1950s. The Marathon III, powered by an AEC 7.7 litre engine and utilizing an AEC gearbox, was their first postwar oil-engined chassis, being preceeded in 1946 by the Mk.II which relied on a 7.4 litre 6-cylinder petrol engine, rather improbably itself derived from a 1928 design.

At their peak - around the date of this photograph - Maudslays manufactured 200 chassis a year, but were, effectively, to become part of AEC in 1948.

The Gloster-Gardner was destined to be one of the rarest coach chassis built. The Gloucester Railway Carriage & Wagon Company found themselves slightly short of work in 1932 and tendered to build a batch of coaches for Red & White. By all accounts the chassis were really well made and this relatively early use of the Gardner 6 LW diesel should have spelled success. Certainly fuel economy was excellent. It may have been that as the engines were rigidly bolted to the chassis vibration was excessive, or, as had been suggested, handling may have been suspect. It could even have been that the drivers' requirements were totally overlooked, as they often were; which understandably could lead to such vehicles being broken with some regularity. No repeat orders followed and, worse, the bodywork, sumptious enough inside, was shortlived only serving for five years when the chassis were rebodied by Duple at an approximate cost of £750 each. The photograph is dated June 1933. Described as a very attractive vehicle "in a masculine sort of way".

Hill's Conducted Tours of Old Meeting Street, West Bromwich, added this diesel engined AEC Regal to their fleet in 1937. Weighing 6 tons 3 cwt (6262 kg) compared with the 8¼ tons of Wrights AEC she was, nevertheless, an excellent example of Duple bodywork. Seating is, of course, luxurious and the irregular line of the bodywork with its paired windows was a fashion in the late 1930s but did not really affect passengers vision. Twin sunroofs were fitted. The curious windowless dummy front was another quirk and one suspects, not one appreciated by a mechanic.

Wright & Sons of Newark-on-Trent are, happily, still trading. Their business was first founded in 1926, when they commenced operating from the Lincoln Road Garage. Initially like many operators they turned their hands to any honest motor trade, including repairs, petrol sales and coach and bus services.

This AEC Regal Mark IV underfloor-engined chassis, with its unusual 41-seater Beccols bodywork was delivered in 1052, the fourth AEC in the fleet. Apart from having one of the most sumptuously seated coaches available, passengers were able to enjoy the use of radio (with twin speakers), an excellent ventilating and heating system for winter and sliding roof panels for summer. Cantrail lights, compulsory from 1992, were fitted 40 years ago!

MIDSCENE - LINK MACHINES

This waggonette was photographed about ninety years ago, and shows the type of vehicle that was hired for days out, weddings or other social occasions. In this case either one of the young ladies was from the (then) village of Yardley, where haymaking and similar agricultural work was the norm, or this outing was actually from somewhere near Yardley. Was there a milliners-cum-drapers in this area? Everyone is very smart and the hats were the epitome of fashion. Shadows would seem to indicate this was a fine, warm, summer's day; in bad weather such an outing, however eagerly looked forward to, became purgatory - but we can only hope!

C 5919 was a Clarkson paraffin-fired steamer of a quite remarkably advanced design, the 'radiator' acting as a condensor and giving a good range per tankful of water.

Originally owned by the rather ephemeral Nantwich & Crewe Motor Bus Company she is seen here taking water from a hydrant at Starbeck, Harrogate, probably in 1911.

The design of the body with longitudinal seating downstairs is as modern as could be found around that time and the original bus route was only unsuccessful due to a lack of passengers. The youthful conductor may be noted, 12 was quite a normal age for their employment, but he is equipped with a form of Bell Punch ticket machine.

It is a straightforward fact of life that if anything can go wrong it will. The North Staffordshire Railway, as proprietors of the Leek & Manifold Light Railway, announced that this line would be opened for traffic on 27 June 1904. They could not have anticipated that the section between Leek and Waterhouses would still be incomplete.

They had, however, decided to maximise revenue and had purchased two Straker steam buses which operated over this part of the line from 23 May 1904. Perforce these steam buses eventually continued to operate until 1 July 1905 when the line finally opened throughout. E223 shows her layout in June 1904. Eventually and frugally, the omnibus shells were dismantled and the two steamers were fitted with pantechnicon bodies.

Over the years the phrase "I've been on a chara trip" has entered British folklore. The pure charabanc with one door to each row of seats was relatively short lived. The whole structure was obviously very weak and as coach chassis were relatively flexible few of these bodies survived for more than three or four years. But the 'charabanc' so loved by people in the 1920s and 1930s usually turned out to be an open top coach with one or two doors, and a lightweight chassis capable of a good turn of speed.

The top photograph is somewhat of a mystery; the vehicle is a Midland Red 'Garford' which were operated between 1922 and 1925. The weather is clearly dreadful and a 'chara' trip on such a day more a matter of survival than pleasure! The oblique parking of the coach for the photographers pleasure says much for the traffic volume that day!

The bottom photograph we do know more about. The coach is probably a Minerva, the location is Ludlow in 1927 and the occasion a workman's outing from the Worcester & Birmingham Canal's workshops at Tardebigge, Worcestershire. The canal superintendant is the one with the beard and buttonhole in the centre and the workshop foreman on the left. Second from the right is the administration clerk. Only one week's paid holiday was granted then, but this outing, eagerly awaited, was partially subsidized. For many of the wives it was 'the' outing, for their lives in some of the canal cottages were terribly lonely, especially as canal traffic declined; cottages could be three or four miles from the nearest road, few had running water or even a well, while in the 1960s one cottage on a Midland canal still relied on a spring ¼-mile away in the fields for its water supply.

This carefully posed photograph shows two London General Omnibus Company's 1920 Type AEC coaches. Built at Walthamstow the bodies were neither char-a-banc nor 'proper' coach styles, access being through a single door on the nearside. The hood, as can be seen, folded forward, the advantage of better reversing ability being claimed but it is hard to see how a driver could cope with this soaking wet combination of wood and canvas situated in front of him. More importantly to the Directors, space (and hence revenue) was lost, these vehicles seating 23. The driver of the foremost, Mr. Clifford was from Darlaston and later returned to the Midlands, ending as Foreman with Midland Red.

AUNTY'S BUS

There is some degree of uncertainty how this photograph got into the hands of its owner. Apparently the story goes that 'Aunty' from Ombersley was given it by a 'chauffeur' or coach-driver with whom she had an understanding and that they had had trips together in the vehicle. What is certain is that it is an AEC Regal model 662 of 1931, powered by a six-cylinder 6.1 litre petrol engine said to have been as smooth as a Rolls Royce. The bodywork is of an odd, thankfully short-lived design by Hall Lewis the coachbuilders. It will be seen the driver has a solid bulkhead behind him and his own rollback canvas roof. Presumably when the vehicle was in the sunshine mode and the rains came down he first stretched the main saloon canvas over the paying customers, then his own bit after which he could settle down into his doubtless soaking wet seat. The underfloor luggage lockers were a rare and advanced fitting, but the emergency door at the rear rather retrograde; if (Heaven forbid) the coach had fallen onto the nearside, the only way out would be via the roof. The legal owner would seem to have been William Baker Neville of London SE 18, and the vehicle's speed restricted to 20 mph. The monogram on the door is indecipherable. There is no registration plate but the Dunlop SS Cord tyres have clearly had a little wear and the mudflaps are hardly new. The lack of curtains tended to indicate this was a successor to the char-a-banc and designed for day trips rather than touring.

Co-operation between road and rail operators became quite 'cosy' in the 1930s as private carriers were eliminated. Both Bristol and Midland Red had strong financial links with the railway (Bristol 50% owned by the GWR and Midland Red; 20% GWR 30% LMSR) and this service originated by the GWR on 1 July 1929 was handed over from 31 January 1932. Apart from providing useful inter-railhead links this combined rail/coach was regularly used by excursionists, thus one could leave Paddington at 9.15 a.m., take the coach via the undeniably scenic route to Burford where, having partaken of lunch one could catch the 2.50 p.m. vehicle to Banbury. A stroll hand-in-hand with the girl around Banbury, catch the 5.15 p.m. and be back in London at the respectable time of 7.28. The fare for the whole coach journey was 4/10d. each, the inclusive rail fare being 12/6d so (with meals) expenditure need not exceed £1.5.0d. (£1.25) each for a glorious day in the country.

"BRISTOL" and MIDLAND "RED" Joint Combined Rail & Road Services.

(IN ASSOCIATION WITH THE GREAT WESTERN RAILWAY)

Commencing MONDAY, September 11th, 1933,
and WEEK-DAYS and SUNDAYS until further notice,
The undermentioned OMNIBUS SERVICES will operate between

SWINDON AND BANBURY

(Via LECHLADE, BURFORD, SHIPTON-UNDER-WYCHWOOD, CHIPPING NORTON, AND BLOXHAM), Connecting with
EXPRESS TRAINS at SWINDON JCT. & BANBURY (G.W.) STATIONS.

		WEEK-DAYS								SUNDAYS				
		A.M.	A.M.	A.M.	P.M.	A.M.	P.M.	P.M.	P.M.	P.M.	A.M.	A.M.	P.M.	P.M.	P.M.
RAIL Paddington dep.		..	..	9 15	..	11c15	..	3 15	..	5 55	..	..	9 10	12 30	..	..	..
Reading		..	..	10 7	..	11W40	..	3 59	..	6R 8	..	..	10 7	12B50	..	..	..
Bristol (Temple Meads) dep.		6 30	9 40	..	12 0	..	3 15	..	6 8	..	..	9 10	..	..	1 30	..	..
Bath		..	9 59	..	12 22	..	3 35	..	6 29	..	..	9 30	..	..	12K53	..	..
Swindon Junction arr.		8 6	10 41	10 58	1 4	1c28	4 16	4 43	7 11	7 21	..	10 15	11 6	2 12	2 20	..	..
		A.M.		A.M.	P.M.		P.M.		P.M.			A.M.		P.M.		P.M.	
ROAD MOTOR Swindon Junction Station dep.		8 22		11 3	1 45		4 48		7 26		..	11 12		2 45		5 45	..
Swindon (Regent Circus) "		8 25		11 6	1 48		4 51		7 29		..	11 15		2 48		5 48	..
Stratton St. Margaret "		8 34		11 15	1 57		5 0		7 33		..	11 24		2 57		5 57	..
Stanton Road "		8 38		11 19	2 1		5 4		7 42		..	11 28		3 1		6 1	..
Highworth "		8 45		11 26	2 8		5 11		7 49		..	11 35		3 8		6 8	..
Inglesham "		8 57		11 38	2 20		5 23		8 1		..	11 47		3 20		6 20	..
Lechlade "		9 0		11 41	2 23		5 26		8A 4		..	11 50		3 23		6 23	..
Filkins (Lamb Inn) "		9 10		11 51	2 33		5 36		—		..	12 0		3 33		6 33	..
Burford (Cotswold Gateway Hotel) "				P.M.								P.M.					
		9 25		12 6	2 48		5 51				..	12 15		3 48		6 48	..
Burford (The Tolsey) arr.		9 27		12 8	2 50		5 53				..	12 17		3 50		6 50	..
Burford (The Tolsey) dep.		9 29		12 8	2 50		5 53				..						
Fulbrook "		9 32		12 11	2 53		5 56				..						
Shipton-under-Wychwood "		9 42		12 21	3 3		6 6				..						
Chipping Norton (Town Hall) "		10 2		12 41	3 23		6 26				..						
South Newington "		10 24		1 3	3 45		6 48				..						
Bloxham (Church) "		10 29		1 9	3 51		6 54				..						
Banbury (Town Hall) "		10 42		1 21	4 3		7 6				..						
Banbury (G.W. Station) arr.		10 43		1 22	4 4		7 7				..						
		A.M.	A.M.	P.M.	P.M.	P.M.	P.M.	P.M.	P.M.								
RAIL Banbury (G.W.) dep.		10 48	10 54	1 25	1 50	4 12	4 26	7 12	7 55		..						
Oxford arr.		..	12 D6	..	2 18	4 40	..	7 43	..		..						
Reading "		..	1Z 0	..	3P 0	6 5	..	8 25	..		..						
Paddington "		..	12 5	..	4V20	7E28	..	9 20	..		..						
Leamington Spa arr.		11 12	..	1 50	..	5 0	..	8 30	..		..						
Birmingham (Snow Hill) "		11 46	..	2 20	..	6T10	..	9T10	..		..						
				A.M.	A.M.	A.M.	A.M.	P.M.	P.M.								
RAIL Birmingham (Snow Hill) dep.		..	..	..	10 0	..	12 45	3 0	..		..						
Leamington Spa "		..	..	..	10 27	..	1 19	3 28	..		..						
Paddington dep.		..	5 30	8 40	11c15	..	..	2Y10	..		..						
Reading "		..	6 18	9 27	12P10	..	..	2 30	..		..						
Oxford "		..	7 20	10 15	12 55	..	..	3 30	..		..						
Banbury (G.W.) arr.		..	8 9	10 43	10 51	1 22	1 46	3 52	4 20		..						
		A.M.		A.M.	A.M.	P.M.		P.M.	P.M.		P.M.	P.M.		P.M.		P.M.	
ROAD MOTOR Banbury (G.W. Station) dep.		..		8 17	11 10	1 56		4 30			..						
Banbury (Town Hall) "		..		8 18	11 11	1 57		4 31			..						
Bloxham (Church) "		..		8 30	11 23	2 9		4 43			..						
South Newington "		..		8 36	11 29	2 15		4 49			..						
Chipping Norton (Town Hall) "		..		8 58	11 51	2 37		5 11			..						
Shipton-under-Wychwood "		..		9 18	12 11	2 57		5 31			..						
Fulbrook "		..		9 28	12 21	3 7		5 41			..						
Burford (The Tolsey) arr.		..		9 31	12 24	3 10		5 44			..						
Burford (The Tolsey) dep.		..		9 33	12 33	3 11		5 44			..	12 25		4 33		7 5	..
Burford (Cotswold Gateway Hotel) "		..		9 35	12 35	3 13		5 46			..	12 27		4 35		7 7	..
Filkins (Lamb Inn) "		..		9 50	12 50	3 28		6 1		8 5	..	12 42		4 50		7 22	..
Lechlade "		..		10 0	1 0	3 38		6 10		8 5	..	12 52		5 0		7 32	..

HAVE COACH, WILL TRAVEL

A country scene can be quite deceptive and in fact this photograph was taken at the back of Cheltenham Coach Station, in the mid 1930s. MT 2048 has a Leyland Tiger TS 2 chassis, and a strangely bus-like body from the same firm. This design of body was not one of Leyland's most successful, rarely lasting more than 8 or 9 years, and here we can see that the side is no longer as straight and true as when she was built. Although severe on the outside the interior was of the highest quality with excellent seats, polished wood, some degree of warmth (although rugs were available), a toilet and gentle lighting for the night. There was, in addition, a chocolate vending machine! Built 1928, sold out of the Red & White fleet in 1937.

The adjustable headlights were an interesting touch, being readily adaptable to weather conditions but one would hazard a guess there was some degree of vibration imparted by those stalks. Cantrail lettering imparted the information that Red & White served South Wales, London, South Coast, Midlands and the North.

The early life of MT 2048 may be found on pp.11 and 12 and it was upon vehicles like this that the whole of Britain's coaching history was built.

JUST PASSING

In July 1929 the following entry appears in the "Official Motor Coach Time-Table":
"Aberystwyth-London via Rhayader, Builth, Leominster, Chipping Norton". The operator was Ensign Motor Coaches, 617 Harrow Road, London W10, the fares 20/- [£1] single and 35/- [£1.75] period return. The service operated daily and took, including meal breaks, 11½ hours, departure being at 0830. A year later a Gilford 1680T coach of Ensign, MY 4210, bearing the notation that it was an "Express Service, London & Cambrian Coast" was photographed in Leominster, having stopped for lunch. The odd items by the radiator are an early form of quasi-independent front suspension using modified Gruss air springs, adjustable according to load to assist the normal semi-elliptic steel springs. The fitting on the bottom of the door is a drop down section which bridged the gap between the door and the step when the vehicle was in motion. Curtains, snazzy light fittings and high level headlights indicate the true longhaul coach. Twenty years later a coach passenger travelled via Associated Motorways, left Victoria at 0845, changed at Cheltenham and arrived in Aberystwyth exactly 12 hours later. The return fare had risen to 37/9d [£1.89].

MOBILE CATERING 1920s STYLE

In the late 1920s, Mr. R. Russell, 'The Pioneer of the Restaurant Car Service' sought to corner, or even make, a 'niche' market, that of providing catering facilities for coach passengers while they were on the move. There were times in the 1950s and 1960s one bitterly regretted that operators had to use some of the ghastly, if not squalid, facilities available. A writer in **Vintage Roadscene** writing about his experiences in the 1950s gave a vivid description of what could be found: ". . . there was another cafe, shunned by drivers except in the case of emergencies and known as 'the fundamental grease-spot'. Had you any sense you took your own eating irons with you as theirs were just literally sloshed through a bucket by Alf as 'Fag-ash Lil' hawking, spitting and blowing ash about her, cooked the food. The daughter, Shirl was equally unsavoury. Without mincing words, when I tell you that Alf always relieved himself on the back wall of the cafe because the septic tank cost money to have pumped out, you will know the sort of place it was . . . [they had steak and kidney pie] . . . and the slower I ate the more the grease separating out from the gravy congealed around the gungy spuds and soggy cabbage. The sight of Shirl trying to be alluring made matters worse - bright lipstick on a sallow spotty skin, lank hair, greasespotted apron over a slightly protruding belly (gained from too many years of leaning on the counter) gave her the charm of a scrap ironing board in a dustcart . . ." But there were not so many cafes in that part of Gloucestershire that a coach-driver had much choice, if a break was insisted upon by passengers who had imbibed too much golden nectar at the seaside.

Mr. Russell's first vehicle was built in 1926, and he experimented with taking 20 friends out for the day and, using the kitchen fitted, he was able to prepare, serve and clear up tea in 20 minutes. On the thoroughly flimsy evidence that: "My friends were so pleased" he was confident "that a restaurant coach service would be a success". After a couple of other 'Restaurant Cars' in 1928 he bought his last and finest vehicle, a three axle Gilford. The snag was that the space lost on catering facilities and their essential adjunct, a toilet, reduced seating to a low level, and it was, of course, necessary to carry a steward. The kitchen measured some 7' (2.3m) long by 4' (1.3m) wide and was fitted with a dresser whereon (shades of 1990s aircraft), all the crockery was kept on small trays "a complete set for each passenger being carried, while stores and glasses wrapped in corrugated cardboard are kept in various drawers". The cooker was "a proper yacht's stove, called a Lathom Cooker" which burned a mixture of paraffin and petrol. There were two large burners plus an oven measuring 12" x 14" (30 x 36 cm) and the use of the whole contraption was "sanctioned by Scotland Yard". Interesting this, as such a combination of fuel and activities added to a petrol engine and its contiguous fuel tank seems rather risky even by 1920s standards. The sink and tap with a pump was supplied by a 10 gallon (45 lt) tank.

Surprisingly, after all the extra expense (£250 on the vehicle cost price, i.e. about an additional 15%-20% on the basic model, plus wages at £2.15 [£2.75] a week plus the loss of income from at least six seats) the menu only consisted of light refreshments. "Orders are taken from the passengers for sandwiches, biscuits, tea, coffee and minerals or other refreshments which are not mentioned on the tariff. Cigarettes can also be obtained, and icecream in summer. The pantry is stocked each morning with home-made sandwiches and cakes, and home-made coffee - not essence - is carried in bottles". The toilet, incidentally, must have been an Achilles Heel for even with modern 'Blue' no 'Elsan' chemical lavatory is very pleasant to sit by after it has been churned about by a moving coach or boat.

Mr. Russell's example was followed by a number of other, optimistic, operators, with J. Mitchell of 40 London Road, Derby, being one of his disciples. "Pride of the Peak" started as a haulage firm, moving on to charabancs particularly Maudslays, which were operated, before 1925, to North Wales. Mr. Mitchell's operations climaxed with this rather elegant '1930 Model' Leyland Tiger fitted with a United of Lowestoft body and delivered early in 1929. This was heavily used for seaside excursions, particularly Skegness, and was an invaluable adjunct to his Saturday evening service from Derby to Ilkeston for dancers at the Premier Dance Hall. So popular was this run of a summer's evening that three or four journeys could be made. We are told that the coach, or coaches, left the Dance Hall at midnight and that included in the return fare of 1/6d (7½p!) was a corned beef sandwich. Alas Mr. Mitchell lost two vehicles (including 'our' CH 8012) in a garage fire during September 1929 and, barely insured, he was forced to give up his operations. Coincidentally, Russell's Restaurant Car Service ceased to trade around the same time although the reasons given for its cessation are conflicting.

THUMBNAIL SKETCH 1 - EVERTONS OF DROITWICH

The predecessors of Everton Coaches, Chapel Bridge, Droitwich, commenced PSV operations in August 1949 and they were able to expand rapidly in the halcyon days of the 1950s and 1960s, absorbing the business of M.E. Ward in 1957 and Williamson of Worcester ten years later. In 1969 they acquired the licences (then often worth more than the coaches) of Crowther of Wychbold. One remembers certain quirks in their services, the whites of a drivers eyes showing on a double-deck football special from Droitwich and Bromsgrove as the youths' 'exuberance' became excessive and the way the passengers waiting to go to Worcester ignored the red and black of all black Everton offerings, stepping back as the driver stopped. So engrained was the 'Midland Red syndrome' that they declined to board even though, often, a coach was offered instead of 'Red' bus. When Evertons were taken over by Hardings of Redditch, their site was built on for residential purposes.

JAB 692 Foden PVSC6/Metalcraft C33F -
operated 3/50-3/62

PWP 200 Commer Avenger III/Duple C41F -
operated 1/56-12/63

UAB 57 Commer Avenger IV/Duple C41F -
operated 1/58-12/63

WUY 692 AEC Reliance 2 MU3RV/Duple C41C -
operated 5/59-2/74

838 CNP AEC Reliance 2MU3RA/Plaxton C41F -
operated 12/60-1/76

NNP 137K Ford R226/Plaxton C53F -
operated 5/72-6/73

TWK 501 AEC Reliance MU3RV/Duple C41F -
operated 5/63-11/66

Positively the last word on a year of jubilees—

London to Cardiff in 1931

by

Today's long-distance express coach
    Is certain to provide
The last word in efficiency—
    A service nationwide.
It wasn't such a rosy picture
    Fifty years ago;
The "good old days" were not so good,
    As these few lines will show.

Let's join the London—Cardiff coach
    Those many years ago.
We ask the driver of the roads:
    "And were they fast or slow?"
"No motorways for us" he says,
    "No rapid cruising speed,
Just narrow twisting stagecoach roads
    That hinder and impede".

"The journey takes eight hours or more
    In this old coach of mine—
From London's cobbled streets we start,
    And though it's not yet nine
There's horse-drawn vans and hackney cabs,
    And many an open bus.
But pressing on, we leave the town,
    Its bustle, noise and fuss".

"Our first stop's Oxford at eleven,
    And at the Cafe Royal
We stop for coffee while I check
    The water and the oil.
At quarter past we're off again,
    And rumble out of town;
It's fifty miles to Gloucester now
    So I keep the throttle down".

"We reach the town in time for lunch
    (I get hungry on these trips)
I'll have to fuel up the coach,
    Then buy some fish and chips.
The garage closed for holidays,
    Now that could spoil my plans,
But I have more luck in Northgate Street—
    It's Pratt's and sold in cans."

"We cross the River Severn,
    Then we're on the road for Ross;
The market's causing quite a jam—
    The cattle have to cross;
We stop again. The radiator
    Shimmers in the sun.
At last we're clear, then into Wales
    To make the homeward run."

"I'll have to stop in Newport—
    There's a parcel in the coach
For the draper in the High Street;
    He waits for our approach—
He wants to keep me talking,
    But I'll have to get away;
There's only twelve more miles to go,
    It's been a long hard day."

"At last we enter Cardiff,
    And draw in to our stand.
A lady sitting at the front
    Slips sixpence in my hand.
"Thank you driver, safely home—
    There's tired you must be!"
Tired? Yes, I'm fit to drop,
    But I'm used to it you see!"

COACHING JOURNAL   DECEMBER 1981

A BRAVE TRY

The London to Cardiff route was one that many coaching entrepreneurs had their eyes on. Thinking ran along the lines of "the Great Western Railway do well, how about us", plus a certain degree of belief that the coach could offer not only better service but a cheaper fare. Routing varied slightly but normally incorporated Oxford and Gloucester, thus following the old A40, but villages like Burford, now almost forgotten, were regarded as good pick-up points. In 1928, Gerald Nowell an ex-London 'pirate' and his wife formed "Great Western Express Company Limited" to provide a fast, comfortable, economic service for the 170 miles (272 km) from London to Cardiff. Initially Tillings-Stevens "Express" chassis were chosen but even with an overall speed limit of 20 mph (32 kph) the "Express" name proved to be a misnomer, and by March 1929 Nowell had purchased three Leyland Tiger TS2 chassis. Time taken was 7½ hours and the return fare was 28/- (£1.40). After a long battle the Great Western Express Company Limited eventually succumbed to competition, initially, from other private operators, but eventually from the big boys of Red & White and Black & White, selling out to the former in 1932. The illustrations on these pages are, however, chosen for their curiosity as much as records of lost coaching activities. The sketch although nominally dated 1931 was of a Studebaker vehicle belonging to Rural England Motor Coaches Ltd who operated between London and Cardiff from April 1927 to January 1930, and I cannot really believe a contemporary driver would have described his vehicle as "this old coach of mine" - such a phrase was, and still is, the quick way to join the dole queue! MT 2049 was one of the GWE Company's Tiger TS2 vehicles, bodied by Christopher Dodson during March 1929. Interestingly at the time trade was so bad for coachbuilders they often took shares in the operator in lieu of cash as part payment, a problem which was to recur in the 1950s and on both occasions was to be the downfall of a number of coachbuilders. But MT 2049 had her body burnt out around August the same year and this photograph shows the rear three quarter view of her new 26-seater body, again by Dodson. The oddity lies in the fact that another, unobtainable for reproduction, photograph shows her registration as being MT 2048. One wonders if some licence chicanery was afoot!

The route taken by 2049 (2048) is clear and of great interest is the method of gaining access to the luggage rack; no boot was fitted and the tarpaulins which covered the rack were far from waterproof. Sadly it must be admitted loading and unloading was normally achieved by one man standing on the roof and throwing down, or catching luggage as required. Bags did not always respond well to this brutal handling!

SWANSEA—LONDON. (Via Cardiff and Oxford).

Read Down	Read Down	Read Down	Towns	Picking-up Points	Read Up	Read Up	Read Up
	9.30 a.m. dep.	9.25 p.m. dep.	SWANSEA	Messrs. Jenkins Motors, Ltd., York St.	7.55 p.m. arr.		9.25 a.m. arr.
	9.40 ,, ,,	9.33 ,, ,,	MORRISTON	Mr. H. Thomas, The Cross	7.45 p.m. dep.		9.17 a.m. dep.
	9.47 ,, ,,	9.41 ,, ,,	SKEWEN	Mr. D. Lloyd, Newsagent	7.38 ,, ,,		9 9 ,, ,,
	9.57 ,, ,,	9.48 ,, ,,	NEATH	Mr. J. B. Dawes, Newsagent, New Street	7.28 ,, ,,		9. 2 ,, ,,
	10. 4 ,, ,,	9.55 ,, ,,	BRITON FERRY	Messrs. D. L. Hancock, The Garage	7.21 ,, ,,		8.55 ,, ,,
	10.10 ,, ,,	10. 1 ,, ,,	ABERAVON	Mr. W. H. Griffiths, 26a High Street	7.15 ,, ,,		8.49 ,, ,,
	10.14 ,, ,,	10. 5 ,, ,,	PORT TALBOT	Messrs. Gaen Bros., Cent. Garage, Talbot Rd	7.11 ,, ,,		8.45 ,, ,,
	10.47 ,, ,,	10.37 ,, ,,	BRIDGEND	Mrs. Frances, Tobacconist, Market Street	6.38 ,, ,,		8.13 ,, ,,
	11. 6 ,, ,,	10.56 ,, ,,	COWBRIDGE	Mr. F. Roberts, 38 Eastgate Street	6.19 ,, ,,		7.54 ,, ,,
9.27 a.m. dep.	11.40 ,, ,,	11.28 ,, ,,	CARDIFF	Red & White Coach Station, Wood Street	5.45 ,, ,,	7.21 p.m. arr.	7.22 ,, ,,
10. 3 ,, ,,	12.18 p.m. dep.	12. 0 p.m. dep.	NEWPORT	Red & White Office, Clarence Place	5.12 ,, ,,	6.45 p.m. dep.	6.50 ,, ,,
10.51 ,, ,,	1. 3 ,, ,,	12.46 ,, ,,	CHEPSTOW	Beaufort Sq., R. & W. Office, High Street	4.27 ,, ,,	5.57 ,, ,,	6. 4 ,, ,,
11.18 ,, ,,	1.30 ,, ,,	1.11 ,, ,,	LYDNEY	Red & White Office, Newerne Street	4. 0 ,, ,,	5.30 ,, ,,	5.39 ,, ,,
11.28 ,, ,,	1.40 ,, ,,	1.19 ,, ,,	BLAKENEY	Mr. Jones, Butcher, High Street	3.50 ,, ,,	5.20 ,, ,,	5.31 ,, ,,
11.38 ,, ,,	1.50 ,, ,,	1.29 ,, ,,	NEWNHAM	Mr. Wheeler, Stationer, High Street	3.40 ,, ,,	5.10 ,, ,,	5.21 ,, ,,
12.15 p.m. arr.	2.30 p.m. arr.	2 0 a.m. arr.	GLOUCESTER	Red & White Coach Station, India Road	3. 0 p.m. dep.	4.30 p.m. dep.	4.50 a.m. dep.
12.45 p.m. dep.	3. 0 p.m. dep.	2.30 a.m. dep.	GLOUCESTER	Red & White Coach Station, India Road	2.30 p.m. dep.	4. 0 p.m. dep.	4.20 a.m. arr.
1.10 ,, ,,	3.25 ,, ,,	2.54 ,, ,,	CHELTENHAM	Clarence Street and Rotunda	2. 6 p.m. dep.	3.35 p.m. dep.	3.56 a.m. dep.
1.47 ,, ,,	4. 2 ,, ,,	3.29 ,, ,,	NORTHLEACH	Mr. Young, Confectioner, High Street	1.26 ,, ,,	2.58 ,, ,,	3.21 ,, ,,
2.12 ,, ,,	4.27 ,, ,,	3.53 ,, ,,	BURFORD	Mr. Packer, Newsagent, High Street	1. 0 ,, ,,	2.33 ,, ,,	2.57 ,, ,,
2.34 ,, ,,	4.49 ,, ,,	4.13 ,, ,,	WITNEY	Mr. Usher, Newsagent, 32 High Street	12.38 ,, ,,	2.11 ,, ,,	2.37 ,, ,,
3. 6 ,, ,,	5.21 ,, ,,	4.43 ,, ,,	OXFORD	Gloucester Green Car Park	12. 5 p.m. dep.	1.39 ,, ,,	2. 7 ,, ,,
3.14 p.m. arr.	5.29 p.m. arr.	4.49 ,, ,,	HEADINGTON	Mr. Pinching, The Shotover Arms Hotel	11.58 a.m. arr.	1.31 p.m. dep.	2 1 ,, ,,
3.29 p.m. dep.	5.44 p.m. dep.	—	HEADINGTON	Mr. Pinching, The Shotover Arms Hotel	11.28 a.m. arr.	1.16 p.m. arr.	—
4.11 ,, ,,	6.26 ,, ,,	5.31 ,, ,,	STOKENCHURCH	King's Arms Hotel	10.48 a.m. dep.	12.34 p.m. dep.	1.19 ,, ,,
4.30 ,, ,,	6.45 ,, ,,	5.48 ,, ,,	HIGH WYCOMBE	Mr. Penn Craft, Florist, High Street	10.18 ,, ,,	12.15 ,, ,,	1. 2 ,, ,,
4.46 ,, ,,	7. 1 ,, ,,	6. 3 ,, ,,	BEACONSFIELD	Beaconsfield Motor Co., Ltd.	10. 0 ,, ,,	11.59 a.m. dep.	12.47 ,, ,,
5. 9 ,, ,,	7.24 ,, ,,	6.24 ,, ,,	UXBRIDGE	Main Road, near Canal Bridge	9.36 ,, ,,	11.36 ,, ,,	12.26 ,, ,,
5.45 ,, ,,	8. 0 ,, ,,	6.57 ,, ,,	SHEPHERDS BUSH	The Lawns	8.45 ,, ,,	11 0 ,, ,,	11.53 p.m. dep
5.55 ,, ,,	8.10 ,, ,,	7. 5 ,, ,,	PADDINGTON	Red & White Coach Stn., Chilworth Mews	8.35 ,, ,,	10.50 ,, ,,	11.45 ,, ,,
6.10 ,, ,,	8.25 ,, ,,	7.15 ,, ,,	VICTORIA	Eccleston Bridge	8.20 ,, ,,	10.35 ,, ,,	11.35 ,, ,,
6.15 p.m. arr.	8.30 p.m. arr.	7.20 a.m. arr.	LONDON	Terminal Station, Clapham	8.15 a.m. dep.	10.30 a.m. dep.	11.30 p.m. dep.

Refreshment stops at Gloucester and Headington on Day Services.

Gloucester only on Night Service.

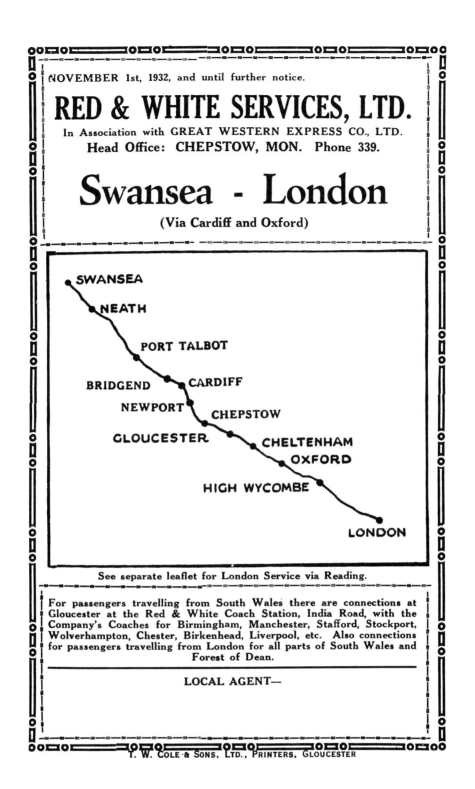

BUS "FANS"

The great interest taken by the rising generation in road vehicles of all kinds - particularly motor buses and coaches - is shown by the huge fan-mail received by many operating companies. The Midland "Red", for instance, daily receives dozens of letters from young enthusiasts asking all kinds of questions about the company, who do their best to satisfy the "fans" demands.

There was, however, a headache or two among the staff of the publicity department when the following request was received from a very youthful admirer:-

"Dere Sir - have you got a Midland Red Bus to spare. I am 9 and allways wanted a Midland Red Bus. I can Learn to drive and will drive for you when I come home from school. I would like a Bus with an upstairs in it and really wheels with big tyres on and a boad with Private on it. Dont bring it till I come home from scholl at 4 o Clock because I shall have to talk to DAD about it. Thank you. Michael T".

The difficulty was overcome by sending Michael some coloured pictures of Midland "Red" buses.

Reported in **Transport World** 8 August 1946

JULY, 1929

ANNOUNCEMENTS OF LONG-DISTANCE SERVICES

ORANGE BROS.,

MARKET PLACE, BEDLINGTON, NORTHUMBERLAND. (Bedlington 36.)
NORTHUMBERLAND HOTEL, EUSTON ROAD, KING'S CROSS, N.W.1. (Terminus 6086 and North 3976.)
(Pioneers of the LONDON and NEWCASTLE Service.)

Pullman Motor Coach Service
LONDON & NEWCASTLE
EDINBURGH & GLASGOW

Bedlington 6.30..Blyth 6.45..Whitley Bay..Tynemouth..North Shields..South Shields..Sunderland (Stothard's) 8 a.m..Durham..Darlington..etc., as per following.

Leave NEWCASTLE, Haymarket, 8 a.m., 8 p.m. *Leave* LONDON, Northumberland Hotel, King's Cross, 8 a.m., 8.30 p.m.

DAILY — Fares — DAILY

	Return.	Single.		Return.	Single.
NEWCASTLE 8.0 a.m., 8.0 p.m.			LONDON (King's Cross) 8 a.m., 8.30 p.m.		
Birtley.			High Barnet.		
Neville's Cross.			Baldock.		
Darlington..	5/6	3/6	Biggleswade..	5/6	3/6
Catterick	8/-	4/6	Alconbury..	12/6	7/6
Leeming Bar.			Stamford..	14/-	8/-
Borough Bridge..	12/6	7/6	Grantham (Lunch)	17/-	9/6
Wetherby..	14/6	8/6	Newark..	18/6	10/6
Aberford.			Retford..	21/6	12/6
Doncaster (Lunch)	17/6	10/-	Doncaster..	24/-	14/-
Bawtry.			Wetherby..	26/6	16/-
Retford..	20/-	12/6	Borough Bridge		
Tuxford.			(Tea)	28/6	16/-
Carlton.			Darlington..	30/-	18/-
Newark..	23/-	14/-	Neville's Cross.		
Grantham (Tea)..	26/-	16/-	Chester-le-Street.		
Stamford..	30/-	18/-	NEWCASTLE..	35/-	£1
Wansford.					
Alconbury.					
Biggleswade.					
Baldock.					
Stevenage.					
Barnet.					
LONDON (King's Cross)	35/	£1			

Through bookings can be made to { EDINBURGH. Fares from London £1.10.0 single; £2.11.6 return. Passengers break the journey at Newcastle, and travel by United and S.M.T. every hour from 7.30 a.m.
{ GLASGOW. Fares from London £1.12.9 single; £2.15.10 return.

BOOKING AGENTS.

LONDON AGENTS.
Head Office, Orange Bros., Northumberland Hotel, Euston Road, N.W.1. Terminus 6086.
Adams & Co., 198, High Road, Wood Green, N.22. and Branches. See p. 54. Palmer's Green 3577.
Australia House Travel Bureau, Australia House, Strand. City 7758.
Orange Bros., 6, York Road, King's Cross North 3976.
Road Travel Bookings Ltd., Bush House, Strand, W.C.2. Central 9534.
Cohen, 363, Green Lanes, Harringay. Tottenham 2363.
Croydon Coaching Centre, 36, High Street, Croydon. Croydon 3235.
Greyhound Motors, Ltd., 229, Hammersmith Road, W.6. Riverside 4273.
Highgate Booking Office, 2, Archway Road, N.19. Mountview 2251.
Claremont Coaching Station, 79-89, Pentonville Road, N.1. Clerkenwell 7343.
Empire Booking Office, 210, Church Street, Kensington, W.8. Park 9393.
Julius & Lockwood, 127, High Street, S.E.13. Lee Green 3566-7.
London Travel Bureau, 409, Holloway Road, N.7. North 3424.
Goodman's, 46, Plumstead Common Rd., Woolwich, S.E.18. Woolwich 1312.
Per Mundum Tour Agency, 207, Camberwell Road, S.E.5. Rodney 4236-7.

Needhams Booking Office, Adj. Chiswick Empire, Chiswick. Chiswick 2109.
Nelsons Tours, Ltd., 8, Grand Hotel Bldgs., Trafalgar Sq., Gerrard 8843.
Rickards, C., 12, Spring Street, Paddington, W.2. Paddington 5686.
Pocock, A. G., 17, High Street, Tally-ho-Corner, North Finchley, N.12. Finchley 3141.
Wheeler, Mrs., 496, Fulham Road, Walham Green. Western 2053.
West London Booking Office, 41, Goldhawk Road, W.12. Riverside 2556.
Victoria Booking Office, 322, Vauxhall Bridge Road, Victoria, S.W.1. Victoria 3811.
Orange Luxury Coaches, Effra Road Brixton. Brixton 4977.
NEWCASTLE.
Miss Vernon, Haymarket, Newcastle Central 5336.
Mason & Co., Gateshead. Tel. 696.
DONCASTER.
Fisher's Corner House Cafe, Waterdale, Doncaster. Tel. 1229.
SOUTHAMPTON.
Kingston & Modern Travel Ltd., 68, St. Mary's Road, Southampton. Tel.: 5902.
CHATHAM.
Orange Coaches, Chatham. Gillingham 5583.

Tickets can be obtained at the Agents only.

Orange Brothers of Bedlington, Northumberland, were the first operator to regularly run the 275 miles (440 km) from Newcastle to London. Commencing in the summer of 1927 with a twice weekly service, by 1929 place departures were twice daily. Demand had grown to such an extent that on the one hand passengers had to be warned that "Long distance coaches, in general, do not ply for hire like the ordinary local bus, and so cannot be hailed and boarded in the street. Tickets have to be booked in advance from the operators of the services or a recognised coach booking agent . . . before the coach leaves its starting point. In every case the possession of a ticket guarantees a seat. In the case of period return ticket it is a strict rule that if the date and time of returning are not stated when the ticket is first booked, notification must be given at the terminal point in advance of the return journey, otherwise the passenger may be refused a seat" but on the other hand not-so fussy predators were gathering to take any 'spare' passengers. Orange Brothers advantage was that, as their depot was about 24 miles (40 km) to the North of Newcastle they could, travelling via Blyth and North Shields, pick up business people and carry them South without their having to change coaches.

Coventry-built Maudslays were obtained in 1929, their solidity contrasting with the relative flimsiness of the Gilfords used for the inaugural runs.

Competition was incredibly fierce, Armstrongs (Majestic), Phillipsons (Stella), Charlton, Cestrian, Taylor (Blue Band) Galleys, National Coachways as well as Orange Brothers all scrambled for bodies on seats from 1928/29.

A number of companies, weakened by poor loadings in winter, succumbed fairly quickly. Orange Brothers fought back, but finally accepted an offer of £72,000 from United Automobile Services and ceased operating in July 1933.

The joint timetable is dated October 1935 and reflects the slacker winter timings. United, Majestic and Phillipsons were shown separately in this leaflet for licensing purposes. Even to the end Bedlington market place (Red Lion) and Bedlington Station (a town in its own right) were still served; interestingly Terrier Coaches ran from Bedlington market until the late 1960s and two other independents still have their base there.

The two close-up illustrations show a representation of the Bedlington Terrier figure once used by Orange Brothers and the seemingly hand-drawn motif used on some United timetables.

Midlands strength - the mighty Maudslay.

MAJESTIC LONDON - RETFORD - CATTERICK
BISHOP AUCKLAND - NEWCASTLE

	am	pm
LONDON (Victoria Coach Station) .. dep.	7 45	7 45
LONDON (King's Cross Coach Station) .. ,,	8 15	8 15
Stevenage (White Lion Hotel) .. .. ,,	9 25	9 25
Baldock (Allnutt's Cafe) .. .. ,,	9 37	9 37
Biggleswade (High Street) .. .. ,,	9 55	9 55
STAMFORD (Market Place) .. .. arr.	11 35	11 35
	pm	am
STAMFORD (Market Place) .. .. dep.	12 5	12 5
Grantham (Main Street) .. .. ,,	12 55	12 55
Newark (Brook's Garage) .. .. ,,	1 30	1 30
Retford (Market Place) .. .. ,,	2 18	2 18
Doncaster (Waterdale) .. .. arr.	3 0	3 0
WETHERBY .. .. .. arr.	4 10	4 10
WETHERBY .. .. .. dep.	4 40	4 40
BOROUGHBRIDGE (Kelly's Cafe) .. arr.	5 10	5 10
BOROUGHBRIDGE (Kelly's Cafe) .. dep.	5 10	5 10
Leeming Bar (R.A.C. Box) .. .. ,,	5 45	5 45
Catterick Bridge (Hotel) .. .. ,,	6 0	6 0
Darlington (Grange Road) .. .. ,,	6 30	6 30
West Auckland (Mill Bank) .. .. ,,	6 57	6 57
Bishop Auckland (Market Place) .. ,,	7 6	7 6
Spennymoor (Cheapside) .. .. ,,	7 19	7 19
Neville's Cross (Hotel) .. .. ,,	7 33	7 33
Durham (Bus Station) .. .. ,,	7 36	7 36
CHESTER-LE-STREET (Bridge) .. ,,	7 52	7 52
Stanley (Front Street) .. .. arr.	—	8 25
Annfield Plain (Co-op.) .. .. ,,	—	8 30
Consett (Market Place) .. .. ,,	—	8 45
Ebchester .. .. .. ,,	—	9 0
Hamsterley Bank Top .. .. ,,	—	9 5
Rowlands Gill .. .. .. ,,	—	9 15
Scotswood Bridge .. .. .. ,,	—	9 25
NEWCASTLE (Marlborough Crescent) .. arr.	8 15	8 15
NEWCASTLE (Haymarket) .. .. ,,	8 30	8 30
NEWCASTLE (Haymarket) .. .. dep.	9 5	9 5
Morpeth (Market Place) .. .. arr.	9 42	9 39
Alnwick (Market Place) .. .. ,,	10 34	10 27
Berwick .. .. .. ,,	11*49	11 37
Dunbar .. .. .. ,,	—	1 3
EDINBURGH (St. Andrew Square) .. ,,	—	2 10
GLASGOW (Buchanan Street Bus Station) ,,	—	4 40

	am	am
GLASGOW (Buchanan Street Bus Station) dep.	—	11 45
		pm
EDINBURGH (St. Andrew's Square) .. ,,	—	2 5
Dunbar .. .. .. ,,	—	3 29
Berwick* .. .. .. ,,	—	4 46
Alnwick (Market Place) .. .. ,,	—	6 1
Morpeth (Market Place) .. .. ,,	6†25	6 53
NEWCASTLE (Haymarket) .. arr.	7†2	7 30
NEWCASTLE (Haymarket) .. dep.	7 30	7 30
NEWCASTLE (Marlborough Crescent) .. ,,	7 45	7 45
Scotswood Bridge .. .. .. ,,	6 50	—
Rowlands Gill .. .. .. ,,	7 0	—
Hamsterley Bank Top .. .. ,,	7 10	—
Ebchester .. .. .. ,,	7 15	—
Consett (Market Place) .. .. ,,	7 30	—
Annfield Plain (Co-op.) .. .. ,,	7 45	—
Stanley (Front Street) .. .. ,,	7 50	—
CHESTER-LE-STREET (Bridge) .. ,,	8 8	8 8
Durham (Bus Station) .. .. ,,	8 24	8 24
Neville's Cross (Hotel) .. .. ,,	8 27	8 27
Spennymoor (Cheapside) .. .. ,,	8 41	8 41
Bishop Auckland (Market Place) .. ,,	8 54	8 54
West Auckland (Mill Bank) .. .. ,,	9 3	9 3
Darlington (Grange Road) .. .. ,,	9 30	9 30
Catterick Bridge (Hotel) .. .. ,,	10 0	10 0
Leeming Bar (R.A.C. Box) .. .. ,,	10 15	10 15
BOROUGHBRIDGE (Kelly's Cafe) .. arr.	10 50	10 50
BOROUGHBRIDGE (Kelly's Cafe) .. dep.	10 50	11 20
WETHERBY .. .. .. arr.	11 15	11 45
WETHERBY .. .. .. dep.	11 45	11 45
	pm	pm
Doncaster (Waterdale) .. .. ,,	12 55	12 55
Retford (Market Place) .. .. ,,	1 57	1 57
Newark (Brook's Garage) .. .. ,,	2 25	2 25
Grantham (Main Street) .. .. ,,	3 0	3 0
STAMFORD (Market Place) .. arr.	3 50	3 50
STAMFORD (Market Place) .. dep.	4 20	4 20
Biggleswade (High Street) .. .. ,,	6 5	6 5
Baldock (Allnutt's Cafe) .. .. ,,	6 23	6 23
Stevenage (White Lion Hotel) .. ,,	6 35	6 35
LONDON (King's Cross Coach Station) .. arr.	7 45	7 45
LONDON (Victoria Coach Station) .. ,,	8 15	8 15

† Sundays excepted.   * Saturdays only.

ADDITIONAL PICKING UP POINTS BETWEEN LONDON AND BALDOCK:—
OLD BALLARDS LANE, FINCHLEY; HATFIELD; VALLEY ROAD, WELWYN.

PHILLIPSONS   LONDON - DURHAM
JARROW - NEWCASTLE

	am	✠ pm
LONDON (Terminal Station, Clapham Road) dep.	7 45	7 0
LONDON (King's Cross Coach Station) .. ,,	8 30	8 0
Stevenage (White Lion Hotel) .. .. ,,	9 44	9 14
BALDOCK (Allnutt's Cafe) .. .. arr.	10 0	9 30
BALDOCK (Allnutt's Cafe) .. .. dep.	10 0	10 0
Biggleswade (High Street) .. .. ,,	10 20	10 20
Buckden .. .. .. ,,	11 0	11 0
Alconbury .. .. .. ,,	11 13	11 13
Wansford .. .. .. ,,	11 47	11 47
Stamford (Market Place) .. .. ,,	12 0	12 0
	pm	am
GRANTHAM (Blue Horse Hotel) .. arr.	12 45	12 45
GRANTHAM (Blue Horse Hotel) .. dep.	1 15	12 45
Newark (Brook's Garage) .. .. ,,	1 45	1 17
Tuxford .. .. .. ,,	2 26	1 56
Retford (Market Place) .. .. ,,	2 46	2 16
DONCASTER (Waterdale) .. .. arr.	3 24	2 54
DONCASTER (Waterdale) .. .. dep.	3 24	3 24
Aberford .. .. .. ,,	4 20	4 20
Wetherby .. .. .. ,,	4 36	4 36
BOROUGHBRIDGE (Three Horse Shoes) .. arr.	5 4	5 4
BOROUGHBRIDGE (Three Horse Shoes) .. dep.	5 34	5 4
Leeming Bar .. .. .. ,,	6 0	5 30
Catterick Village.. .. .. ,,	6 12	5 42
DARLINGTON (Grange Road) .. .. ,,	6 54	6 24
Durham (Market Place) .. .. ,,	7 31	7 1
Houghton-le-Spring (Sunderland Street) .. ,,	7 45	7 15
Sunderland (Maritime Street) .. .. ,,	8 0	7 30
South Shields (55, Fowler Street) .. .. ,,	8 20	7 50
Jarrow (Monkton Road) .. .. ,,	8 35	8 5
Hebburn .. .. .. ,,	8 40	8 10
NEWCASTLE (Marlborough Crescent) .. arr.	8 45	8 15

	am	✠ pm
NEWCASTLE (Marlborough Crescent) .. dep.	8 0	8 0
Hebburn .. .. .. ,,	8 10	8 10
Jarrow (Monkton Road) .. .. ,,	8 20	8 20
South Shields (55, Fowler Street) .. ,,	8 35	8 35
Sunderland (Maritime Street) .. .. ,,	8 55	8 55
Houghton-le-Spring (Sunderland Street) .. ,,	9 10	9 10
Durham (Market Place) .. .. ,,	9 25	9 25
DARLINGTON (Grange Road) .. .. ,,	10 10	10 10
Catterick Village.. .. .. ,,	10 37	10 37
Leeming Bar .. .. .. ,,	10 50	10 50
BOROUGHBRIDGE (Three Horse Shoes) .. arr.	11 30	11 30
BOROUGHBRIDGE (Three Horse Shoes) .. dep.	12 0	11 30
	pm	am
Wetherby .. .. .. ,,	12 30	12 0
Aberford .. .. .. ,,	12 50	12 20
DONCASTER (Waterdale) .. .. arr.	1 40	1 10
DONCASTER (Waterdale) .. .. dep.	1 40	1 40
Retford (Market Place).. .. .. ,,	2 20	2 20
Tuxford .. .. .. ,,	2 35	2 35
Newark (Brook's Garage) .. .. ,,	3 3	3 3
GRANTHAM (Blue Horse Hotel) .. arr.	3 35	3 35
GRANTHAM (Blue Horse Hotel) .. dep.	4 5	4 5
Stamford (Market Place) .. .. ,,	4 50	4 50
Wansford .. .. .. ,,	5 3	5 3
Alconbury .. .. .. ,,	5 47	5 47
Buckden .. .. .. ,,	6 0	6 0
Biggleswade (High Street) .. .. ,,	6 40	6 40
Baldock (Allnutt's Cafe) .. .. ,,	7 0	7 0
Stevenage (White Lion Hotel) .. ,,	7 16	7 16
LONDON (King's Cross Coach Station) .. arr.	8 30	8 30
LONDON (Terminal Station, Clapham Road) ,,	9 15	9 15

✠ DURING THE PERIODS NOVEMBER 1st TO DECEMBER 15th (INCLUSIVE) AND JANAUARY 14th TO MARCH 31ST (INCLUSIVE) THE MORNING SERVICE ONLY WILL OPERATE.

ROUTE No. 1.
JOHN BULL TIGER EXPRESS SERVICE
(Proprietors: WOOD BROS. (Blackpool) LTD.),
WALTER WOOD, Secretary.

DAILY AT 8-30 a.m., FROM
LONDON to BLACKPOOL
Via COVENTRY, BIRMINGHAM, STAFFORD, BOLTON, DARWEN, BLACKBURN and PRESTON.

FARE LIST.

	Single	Return
LONDON to A GROUP	10/-	15/-
LONDON to B GROUP	12/6	20/-
LONDON to C GROUP	15/-	25/-

CHILDREN UNDER 14, TWO-THIRDS FARE.

	LONDON	dep. 8-30 a.m.—Motor Coach Bookings, Bush House	arr. 7-30 p.m.
	ST. ALBANS	9-22 a.m.—Thos. Hansell, 41, London Road	6-30 p.m.
	DUNSTABLE	9-55 a.m.—The Clock — Tea	dep. 6- 0 p.m. / arr. 5-30 p.m.
A	STONEY STRATFORD	10-40 a.m.—Johnson, 84a, High Street	4-45 p.m.
	TOWCESTER	11- 2 a.m.—Post Office	4-23 p.m.
	DAVENTRY	11-32 a.m.—New Arterial Road	3-40 p.m.
	COVENTRY Lunch	arr. 12-20 p.m. / dep. 1- 0 p.m. — Greyfriar's Green (Outward) Cheylesmere Park (Return)	3- 0 p.m.
	BIRMINGHAM	1-45 p.m.—Smithfield Garage Coaching Station, Digbeth, Birmingham	2-15 p.m.
	WALSALL	2- 8 p.m.—Arboretum Garage, Denmark Road	1-55 p.m.
	CANNOCK	2-18 p.m.—Royal Oak Hotel	1-30 p.m.
B	STAFFORD	2-50 p.m.—Direct Coal Supply, Bridge Street	1-10 p.m.
	NEWCASTLE-U-LYME	3-30 p.m.—Shaw's, Market Place — Lunch	dep. 12-30 p.m. / arr. 11-50 a.m.
	KNUTSFORD Tea	arr. 4-35 p.m. / dep. 5- 5 p.m. — Canute Cafe, The Square	10-45 a.m.
	BOLTON	5-50 p.m.—Christie's Garage, Victoria Square	9-45 a.m.
	DARWEN	6-15 p.m.—Station Garage	9-30 a.m.
C	BLACKBURN	6-30 p.m.—Golden Lion Hotel Yard	9-15 a.m.
	PRESTON	6-55 p.m.—Hesketh's, Myerscough Buildings, Church Street	8-45 a.m.
	"	7- 0 p.m.—Starchhouse Square	8-40 a.m.
	BLACKPOOL	arr. 7-40 p.m.—Coronation Street Garage	dep. 8- 0 a.m.

Times for Intermediate Stages are only approximate.
Drivers must not under any circumstances exceed 30 miles per hour on any section of this Service.

VICTORIA BOOKING OFFICE,
AT THE TRAM TERMINUS,
322, Vauxhall Bridge Road, S.W.1.

"Gazette & Herald," Blackpool.

TIGER! TIGER! BURNING BRIGHT....

Wood Brothers (Blackpool) Ltd., first operated their John Bull 'Tiger' service in 1928 but it only ran until early 1933, when W.C. Standerwick acquired both the route and Wood's Blackpool excursions. Standerwicks were a subsidiary of Ribble and thus Woods ended up being 50% owned by a railway (LMS), the very company that they had fought against.

In the meantime Wood's had purchased some of the most delightful vehicles for the service and those Leyland Tigers headlights did indeed burn bright as they pounded through the winters' evenings with metronomic regularity. The bodywork on this 'Tiger' was by H.V. Burlingham and of a true 'all-weather' pattern. The front bulkhead, roof hoop-sticks to carry the canvas, and the entire rear of the vehicle housing toilet and hand basin were fixed items, but in clement weather the roof could be rolled back. Down the centre of this was an upholstered plank which carried light fittings in addition to the individual reading lights alongside each seat. Seats were covered in leather, albeit somewhat low-backed and the twenty passengers each had a folding table.

It will be seen from the timetable that Woods regarded London-Birmingham as a lucrative source of traffic in its own right; but it would be interesting to know when Greyfriars and The Canute Cafe ceased to be recommended stops on the Blackpool run.

On this vehicle, the first of its kind, the 'Tiger' motif is prominent, later a bulldog coupled to 'John Bull' provided the relief on Woods' yellow livery. The waistband was a deepish red, as were the mudguards. The running board continuing along the full length of the vehicle was a curious throw-back to 'chara' days.

NOT VERY GOOD
For Long-distance Coaching

Sir,

On a recent Sunday I booked a seat to travel from the Midlands to Blackpool by coach. Such a mode of travel would have been very preferable to the rail, which would have taken more than two hours longer. As far as Manchester the journey was really comfortable, but the proprietors of the service, on finding that there were only five passengers for the remainder of the journey from Manchester to Blackpool, decided not to run the coach farther, and they therefore substituted a private car of uncertain age to complete the last stage. There was really room for only three passengers in the back, and I asked the driver if I should sit beside him at the front, but he replied that he was expecting a lady passenger to occupy that seat. The whole five of us, therefore, squeezed into the back of the car, three on the seat and two on a box facing backwards, where we remained for a considerable period, with cramped legs watching the driver speaking to numerous girls passing by in an obvious endeavour to take one of them for a joy ride to Blackpool and back. Being unsuccessful, he at last invited one of us to sit beside him, and conditions in our compartment were somewhat relieved. Corners were taken at such speed that the car's wings grated on the tyres and groans came from all parts of the car.

Surely conditions different from these should be available to long-distance travellers.

[Signed] A.L. Coventry   July 1929.

In **Bus & Coach** April 1929 an investigative reporter wrote that:
"I was wondering the other day what long-distance coach passengers really think about unheated coaches during bitterly cold winter weather. Candidly, I expected some pretty crisp complaints to be registered at the big coach station I visited in order to get opinions. However, unprintable comments were by no means the order of the day, but there were a number of minor criticisms worthy of serious consideration. The passengers questioned had travelled approximately two hundred miles and the day had been extremely cold. They could be conveniently divided into two categories - those who had felt rather cold but did not particularly object to it, and those who considered that while the provision of suitable rugs was an appreciable addition to comfort it was definitely not adequate in the circumstances. An examination of the rugs used for this purpose on a Manchester-London coach showed that they were the usual plaid type and large enough thoroughly to cover an average-sized person from waist to ankles".

IMPERIAL MOTOR SERVICES
308, UPPER PARLIAMENT STREET, LIVERPOOL.

BEST IN THE LONG RUN

DAILY SERVICE BETWEEN

SINGLE 17/6   **LONDON & LIVERPOOL**   **RETURN 30/-**

via UXBRIDGE, HIGH WYCOMBE, OXFORD, STRATFORD-ON-AVON
WARWICK, KENILWORTH, LICHFIELD, STONE, KNUTSFORD, WARRINGTON

VIA THE HEART OF ENGLAND

Until 1935 the passenger intending to travel from Liverpool to London had an extensive (if declining) choice of routes. In 1929, for example, he or she could, if they were so minded route themselves via Chester, Newport, Coventry and Daventry on Samuelson's Saloon Coaches, the successor to the long established Samuelson Transport Company. Samuelsons also offered an alternative daily service via Leamington and Warwick. Both took 10½ hours and cost 75 pence single. Via Chester, Newport, Stratford and Oxford the roads were positively crowded with Rymers Tours and MacShanes Motors, offering a total of seven daily departures. Should you wish to travel via Chester, Birmingham and Oxford this was feasible (but subject to load) whereas Imperial Motor Services (already financially in trouble) would be pleased to accept your 75p and carry you via Warrington, Lichfield and Stratford. Crosville, not then a monopolist, ran via Crewe and Coventry taking eleven hours, while delightfully, Albatross Roadways Ltd., routed via Manchester and not stopping took 9 hours but charged £1.25. Theirs was a sleeper service of the highest quality.

Two companies are missing from this list, and both were unique in their time. Motorways of Haymarket, London, were founded in 1920 to provide continental touring facilities, reasoning rightly that not only would there be a desire to see Mam'selle from Armontieres but that parents would wish to visit where their sons fought. Few, very few, ordinary people had ventured abroad pre-1914, and an organised tour by motor-coach which promised luxury, buffet, toilet and an attendant while ridding one of the hassle that was (and is) concomitant upon foreign travel was to be welcomed. Conversely, tourists visiting Britain also desired an easy entry and this Motorways promised them, as can be seen from this re-produced brochure. "Motor Pullmans will leave by road . . . in connection with the Arrival and Departure of the Principal Atlantic Steamships". In the summer of 1927 this meant three arrivals at Liverpool in May, four in June and July, five in August and four again in September. Twenty ocean liners of the highest class in five months! Would that we could recapture "The Old Country as she was", Birmingham, Warwick, Stratford, each seemingly charming places. The Pullman is believed to be a new Duple bodied Berliet, seating 12 with a buffet, attendant and driver. Presumably the driver (or, indeed, chauffeur) had some way of reducing the rain's downfall he was bound to drive through. Motorways for a number of reasons failed in 1929, were reformed in 1930 but somehow lost their zest.

A complete contrast with the vehicle of Motorways (and, alas, making it seem outdated) is the 1932 Duple bodied AEC Regal of Pearson's Motorways which was one of the first built to the new overall permissible length of 27'6" (18'6" wheelbase). Still petrol engined at about 6-7 m.p.g. this machine was fast for a heavyweight and much more solid in its ride than the Gilfords and continental lightweights that were often to be found. The louvres above the windows give Pearsons chosen route, the nearside shows Liverpool, Birkenhead, Chester, Oxford and London, and the readable glasses on the offside have the addition of Stratford-on-Avon. Everything about the vehicle depicts its newness and, indeed, it is not yet registered.

By 25 May 1935 J. Pearson and Sons, Happy Days Motorways, were to have the unenviable distinction of being the last of the Independents, handing their fleet over to a Crosville/Ribble consortium on that date.

Imperial, whose leaflet is reproduced, were said to be the first operator to run daily and to complete the journey in the day (Motorways took two days) commencing on 2 May 1928. Although departure ex London (27 Cartwright Gardens W.C.2) is given as 8.30 a.m., no arrival time is even hinted at.

THE MOTORWAYS PULLMAN
at the Liverpool Landing Stage

YOU land at Liverpool. The advantage is yours, for from this port you can save two days of your holiday by taking in many places of interest on the way to the Capital. There, right at the landing stage, awaits a luxurious Motor Pullman, ready to take you straight away to London through the heart of England. The Motorways Pullman, built on the lines of the fastest and most richly equipped American automobile, takes a road that leads through the wonderful Shakespearean country to Oxford and the pleasant valley of the Thames. Your armchair is ready for you. Your luggage is stowed safely away. The uniformed courier gives the signal, and the powerful car glides forward swiftly and silently. You are entering England.

It is an experience that you will never forget. As you pass along you will see the Old Country as she really is, in her ordinary natural garb. The road winds on through sleepy villages and old-fashioned market towns, by rustic by-ways and mellowed farm houses, now through a busy town thriving in the shadow of the old Cathedral, now past a lonely hamlet nestling in smooth folded hills. Little by little you will sense the spirit of antiquity that walks hand in hand with the spirit of to-day. You will catch glimpses of the richly coloured pageant that has passed along the broad, white English roads.

The ancient walls of Chester—the busy hum of Birmingham—the mediæval charm of Warwick—the fragrant memories of Stratford-on-Avon—the silver ribbon of the Thames. These are the milestones on your way to London. When the time comes for you to turn homewards once more, and when you have "done Europe," why not finish up your holiday with a last memory of rustic England? The road turns north through Buckinghamshire, past the quaint but charming villages of the Penn and Washington countries, so rich in American association, and again to Stratford-on-Avon, this time following the road through Bridgnorth and Much Wenlock to Chester. Thus your picture of England is complete, and you have wasted neither time nor energy by your decision to

"See England Right Away"

ANN HATHAWAY'S COTTAGE

THE ROAD TO LONDON
First day. LIVERPOOL—Chester. Nantwich—Stone. Lichfield—Birmingham. Dinner. Night. Breakfast at the Midland Hotel.
Second day. Kenilworth—Warwick. Leamington—Stratford-on-Avon. Lunch. Shipston—Chipping Norton. Woodstock—Oxford. Maidenhead. LONDON.

THE ROAD TO LIVERPOOL
First day. LONDON—Beaconsfield. Aylesbury—Buckingham—Winslow. Lunch. WASHINGTON COUNTRY. Sulgrave—Banbury. STRATFORD-ON-AVON. Dinner. Night. Breakfast at the Red Horse and Golden Lion Hotel.
Second day. Worcester—Kidderminster. Bridge Norton—Much Wenlock. Shrewsbury. Stop for lunch. Whitchurch—Chester. LIVERPOOL.

The two journeys are thus both varied and complementary. The single fare in either direction is 20 dollars, or £4/2/6. The Return fare 39 dollars, or £8. Including:—
The Motor Tour.
1st class hotel accommodation and meals.
Free carriage of a reasonable amount of heavy baggage from Liverpool to London hotel or vice versa.
Services of Courier-Lecturer.
All gratuities.

Passengers are requested to limit their hand luggage in the Pullman to one suitcase not exceeding the following dimensions—24 × 17 × 7.

A TYPICAL ENGLISH COUNTRY LANE

WADHAM COLLEGE, OXFORD

"See England Right Away"

19

MKV 6 Daimler D650HS/Burlingham C39C
operated 7/53 - 7/63

BWK 48 Maudslay SF40/Duple C32F
operated 7/36 - ?/40

CDU 447B Ford 676E/Duple C252F
operated 7/64 - 4/70

GDU 7 Leyland PS 1-1/Burlingham C33F
operated 7/47 - 10/52

KRW 608E AEC Reliance 6U3ZR/Duple C49F
operated 7/67 - 4/71

MKV 1 Bedford SB/Plaxton C35F
operated 6/53 - 5/60

THUMBNAIL SKETCH 2 - RHMS, COVENTRY

The Red House Garage Company (later Red House Motor Services) was founded in 1919, as a private venture, became a Limited Company in 1926, and grew by absorption, growth enhanced by the fact that the companies they took over had, themselves, in earlier days bought up the licences of a number of other Coventry based independents. Owned by the original family from 1919 to 1977, their bus (stage carriage) routes were sold to Midland Red in 1936, and their strength lay in the coach traffic which took vehicles to destinations as far apart as Margate, Blackpool, Scarborough and Great Yarmouth. Two characteristics of the firm were a liking for 'different' vehicles and that as shown in the photographs of KHP 895E and KRW 608E rather than letter their coaches under separate names, all the absorbed firms are shown on the sides of the vehicles: Bantam, Godiva, B.T.S., Bunty and RHMS.

KHP 895E Albion VK 43L/Park Royal C43F
operated 5/67 - 6/76

THE ROLE OF THE TRAFFIC COMMISSIONER
By J MERVYN PUGH, TRAFFIC COMMISSIONER FOR WEST MIDLAND & SOUTH WALES TRAFFIC AREAS

England, Scotland and Wales are divided into eight Traffic Areas and there are seven Traffic Commissioners, the Commissioner for West Midlands also being responsible for South Wales. Commissioners are also the Licensing Authority for goods vehicles and act as agents for the Secretary of State for Transport in matters relating to the conduct of drivers who hold, or wish to hold, a vocational entitlement (PCV and LGV).

Each Commissioner has a deputy Commissioner or Commissioners. They are generally lawyers, but not always, and they sit as and when required. Each year there is a Commissioners' Training Seminar, chaired by the Senior Traffic Commissioner, to which all Commissioners and deputies attend. Deputy Commissioners are not involved in the day to day running of the Traffic Area office, nor are they concerned with policy decisions: these are the responsibility of the Commissioner.

A Senior Traffic Commissioner is appointed from the ranks of the Commissioners by the Secretary of State for Transport. He has no command lead over individual Commissioners and their deputies as to how they conduct themselves but his role is one of advising as to office procedures and general policy in an effort to obtain consistency of approach throughout the eight Traffic Areas.

Each Commissioner is independent of the Department of Transport, the Secretary of State and Ministers, and thus it follows independent of each other. Commissioners, despite this independency, do meet regularly and discuss all matters appertaining to their role and try to be consistent each with the other not only as to office practice but procedural matters at Public Inquiries.

The Commissioner, besides his responsibilities for licensing and disciplining of operators as is mentioned hereafter, is also involved in attending trade association meetings and liaison with the Bus and Coach Council is essential for the smooth running of a Traffic Area. Trade unions too are involved and close co-operation with them in particular with drivers' conduct is all important. The support that the trade unions give to Commissioners is superb and praise and thanks must go to them, in particular the Transport and General Workers' Union.

The focus of the Commissioner's responsibilities, in relation to Public Service Vehicles, has changed from Road Service Licensing to Safety. The Transport Act of 1985 de-regulated local bus services, thus enabling any operator to run any route of his choice with no restriction upon the amount he charges for fares, provided that the local service is registered with the office of the Commissioner whose responsibility is then to see that the operator operates fairly and keeps to the timetable which he lodges with the Commissioner when he registers the local service.

Each and every passenger carrying vehicle (PCV), which are still referred to as public service vehicles in the Statutes [The Public Passenger Vehicles Act 1981 and the Transport Act 1985] and Regulations, which has eight or more seats and is used for the carriage of passengers for hire or reward, has to be the subject of a PSV operator's licence. PSVs may not necessarily have eight passenger seats: a number of cars are subject to an operator's licence. A vehicle operated as a PSV may not necessarily be of the construction traditionally associated with PSVs but may be any vehicle that is operated under a PSV licence. The licence is held in the name of the operator who applies for a public service vehicle operator's licence to the Commissioner in whose traffic area he has his operating centre. There are two main grades of licence, the Standard National licence, which is sub-divided to cover International Standard licences, and the Restricted operator's licence which is a licence held by an operator whose main employment is outside the industry.

Commissioners may also issue bus permits which enable voluntary groups and certain other bodies to make a charge for providing transport to their own members and others whom the organisation serves without having to comply with the full PSV licensing requirements.

Prospective operators have to satisfy the Commissioner as to their financial standing and good repute, and that they will maintain their vehicles in a fit and serviceable condition. Good repute is a matter of fact but can relate to the previous holding of an operator's licence and to any offences committed relating to public service vehicles. Notice of each application for a PSV licence is published fortnightly in the departmental publication 'Notices and Proceedings', which is circulated in all local authorities and police forces in each traffic area who may object to the grant of a licence on the grounds that one or all of these criteria are not met.

Operator licences are granted by the Commissioner in whose area the operating centre is and have effect anywhere in Great Britain. An operator may hold more than one operator's licence but not more than one in each Traffic Area.

An application for a Public Service Vehicle operator's licence, which is generally granted (provided that good repute, financial standing and maintenance arrangements are satisfactory) for a period of five years in the West Midland and South Wales Traffic Areas heard at a Public Inquiry. At the time of grant the new operator has explained to him his responsibilities and is then further invited to attend a New Operators' Seminar and to bring with him other persons associated with his business. At these Seminars, besides the Commissioner attending, senior representatives of the Vehicle Examiner and Traffic Examiner departments of the Vehicle Inspectorate also attend and, after the Commissioner has departed, conduct a question and answer session. These Seminars have proved to be invaluable.

There is no such licence application as a renewal: when a licence expires a new application must be made and if such application is made prior to the expiration of the existing licence the operator may continue under his 'old' licence prior to his new application being considered and the licence granted. [Continuous licensing is, at the time of writing, being seriously considered, as are the imposition of environmental controls on operators' centres similar to those already in existence relating to goods vehicles.]

On applying for a Public Service Vehicle operator's licence each applicant signs a declaration that the statements made in this application are true and then makes a 'statement of intent' by signing the following assurances:-

"I will make proper arrangements to ensure that:
* the laws relating to the driving and operation of vehicles used under this licence are observed:
* the rules on drivers' hours are observed and proper records kept;
* vehicles do not carry more than the permitted number of passengers;
* vehicles are kept in a fit and serviceable condition;
* drivers report mechanical faults in vehicles as soon as possible; and
* records are kept (for 15 months) of all safety inspections, routine maintenance and repairs to vehicles, and made available on request.

I will
* notify the Traffic Commissioner of any changes or convictions which affect the licence; and
* maintain adequate finance resources for the administration of the business."

It is safety though that is paramount and each and every operator has to satisfy the Commissioner that he has a system of preventative maintenance. The Commissioner, in being satisfied, takes into account the number of vehicles on the licence, the anticipated mileage and general use of vehicle. Whilst it is not the Commissioner's duty to set out guidelines which must be adhered to, his role has, during the past few years, become far more educative than ever before and thus it follows that Commissioners do give advice. It is of course up to the operator as to whether or not he accepts that advice but most operators do take on board advice given by a Commissioner in that the operator is well aware of the fact that the advice given can, if heeded, prevent prohibition notices being placed upon his vehicles.

A few years ago maintenance was carried out, generally speaking, upon achieving a time limit and/or mileage alternatives, thus 6,000 miles or 3 months whichever first shall happen could have been the period put forward. Now the mileage alternative has passed into history and all inspections should be time based. All operations must revolve around the maintenance and when there was a mileage alternative it so often happened that an inspection was missed or delayed in that the operator did not exactly know when the inspection would be due and if he was going to an outside garage, as many smaller operators do, the garage frequently was unable to inspect the vehicle when it was due owing to its commitments. Under a time based system of planned maintenance the vehicles are pre-booked in at the garage so not only does the operator know the date and time of the inspection, so does the garage. Similarly with operators having their own fitters, the fitters are now able to work to a regular timetable thus time is pre-allotted on a regular basis for each and every vehicle. A flow chart is recommended and it is suggested that this should be on the Transport Manager or Operator's wall and that colour coding should be used. Thus the inspections flow ahead so that anyone at a glance can immediately see when the next inspection is due. A flow chart, which is in effect a forward planner or diary, can also be put to further use such as retrospectively marking the services and oil changes, the replacement of tyres, and forward dating the MOT and calibration of tachographs. It is recommended that the flow chart should work for the operator and that he should use it in every way that he can. It is even suggested that he should put his wife's birthday on it too!

Maintenance inspections cannot satisfactorily be carried out unless the fitter has in his possession the PSV Inspection Manual which is available from all testing stations and HMSO. This book gives the method of inspection and the reasons for rejection and the inspection sheet has on it manual numbers which coincide with the numbers in the book: thus each operator is able to check from his inspection sheet any defect that is found. To help the fitters the Vehicle Inspectorate have recently published two further books, 'The Public Service Vehicles Categorisation of Defects' and 'The Public Service Vehicle Inspection Manual Amplification Notes'. These are essential books for any person who is responsible for the maintenance of a public service vehicle. They are the old 'Secret' books of the Vehicle Inspectorate which, following the formation of the Inspectorate as a Department of Transport Government Agency, were published. These books too are available at testing stations and HMSO.

Operators who choose to have their vehicles inspected at an outside garage are recommended to visit that garage with their own copy of the Inspection Manual and to watch a fitter inspect his vehicle. This is not to make the operator into a mechanic but to familiarise him with an inspection and in particular with defects that can arise on his vehicle. Sadly it is surprising how many outside garages do not have their own copy of the Inspection Manual, nor the Categorisation and Amplification books. Operators can contract out their maintenance but they cannot contract out their responsibility and thus it is recommended that each and every operator should make certain that if he goes to an outside garage that garage does have the necessary books, and if he employs his own fitters that they too should have their own copies. It is of little help to have the books in the Manager's office. Each fitter should have his own Inspection Manual though perhaps the foreman only in a large garage should have the Categorisation and Amplification books. In a perfect world perhaps each fitter would have a complete set. When a foreman fitter is employed he should be encouraged to master the books and have regular meetings with his fitters to discuss defects. National and International operator licence holders must employ a Transport Manager who must be the holder of a certificate of professional competence. This is acquired by taking an examination set by the Royal Society of Arts. This requirement became law on 1 January 1978 but it was possible for those previously employed or involved in this industry to claim 'grandfather rights' - these gentlemen and a few ladies are now a dying race. The Transport Manager may be a part-time employee and this brings the question as to how many operators he should work for and how many vehicles he should have under his command. He must have a contract of employment with each and every employer and the Commissioner must know who he is and be satisfied both as to his competence and employment. There are five matters for the Commissioner to consider in satisfying himself that the Transport Manager does not have too great a workload. They are:

1. How many and what hours does the Transport Manager work?
2. How many employers has he?
3. How many vehicles does each of his employers have?
4. What is the distance between each of the operating centres for which he has responsibility?
5. What are his duties - do they include being a driver and/or a fitter?

Thus there is no hard and fast rule as to how many operators should employ him or how many vehicles he should have under his command: it is a matter of fact as to whether or not the Transport Manager's workload is acceptable.

Daily 'Nil' reporting of defects is vital to any system of planned maintenance. History has shown that giving drivers 'Tick Boxes' to complete at the end of a shift, or just reporting defects, is not infallible and over the last few years operators have been encouraged in the West Midland and South Wales Traffic Areas to give each and every driver his own personal blank duplicate book and at the end of each shift the driver writes the name or the number of the vehicle, the date and, if there are no defects, he writes 'Nil' and signs it. The top copy is then torn out and handed to the Transport Manager who is recommended to keep the Nil sheets for approximately 21 days. The carbon copy in the duplicate book becomes part of the documentation which must be retained for at least 15 months. Where a defect is recorded it is recommended that the remedial action taken should be written on the report and also that a repair book should be kept. This has in fact proved a money saving piece of advice in that at the end of each financial year the operator's accountant will require from him details of expenditure so that he can claim the necessary tax relief. The handing of a bundle of receipts to an accountant can involve the operator in considerable expense, but by keeping a repair book and photocopying that book the accountant's work is done for him and money is saved.

It is recommended that each operator should write a letter to each driver setting out his daily tasks, for instance a walk round the vehicle checking that the lights, tyres, wheels fittings, bodywork are safe and where appropriate in working order. The full responsibilities of the operator are contained in the 'Guide to Maintaining Roadworthiness' which was produced by the Department of Transport in partnership with the transport industry in 1991. The letter should also include the duplicate book with instructions on how to complete it and it is important that the letter should be in duplicate and that the driver should sign the second copy as being received, read and understood. Where fitters are employed by the operator it is similarly suggested that letters in duplicate should be given to the fitters and that these letters should set out their responsibilities and in particular to stress the safety aspect of their work.

The naming of vehicles is important and, of late, more and more vehicles are baring independent names. This not only brings fun into the maintenance but causes family competitiveness in a small operator which can bring nothing but good. It is not unknown for husband operators to name vehicles after their wives and children and anyone who has a vehicle named after them will tell you that their first priority will be to see that that vehicle is always in prime condition.

This simple system of planned maintenance will only be totally effective if the vehicles are in A1 condition at the commencement and it is therefore suggested that any operator adopting this system should, regardless of any life in his current MOT, have his vehicle newly MOT'd. An MOT speaks only of the day of the test and is no indication as to its future roadworthiness.

Prohibition notices placed on vehicles by the Vehicle Inspectorate can be either immediate or delayed and can be placed on a vehicle at a roadside check carried out by the Vehicle Inspectorate which will be police-assisted, or when a vehicle is presented for a MOT which fails and when the Inspectorate visits the operator's premises. [The police now have similar powers of prohibition.]

Commissioners in the main are concerned with immediate prohibitions which are indicative of neglect and which constitute in themselves a danger to road users. Neglect comes about in three ways: if an inspection is missed or delayed; or more often when it has been a poor inspection.

The Commissioner has wide powers under Section 17 of the Public Passenger Vehicles Act 1981 and a licence can be revoked, suspended or prematurely curtailed and the authorisation can be reduced. These punitive powers are essential if the public are not to be deprived of their assumption that every bus that they get onto at any time will always be safe and safely driven.

Passenger transport is a very competitive area and presentation of both vehicle and drivers is important. Operators are therefore encouraged to have good, smart liveries for their vehicles and uniforms for their drivers. The Commissioner for the West Midland and South Wales Traffic Areas also advocates the retention of the driver's badge and that operators should make their buses 'No Smoking' buses. All London buses are now 'No Smoking' buses and this principle has now been accepted by the industry. It is, of course, an offence for a driver to smoke when passengers are on the vehicle.

Once an operator has obtained his licence, a vehicle identity disc will be issued to him for each vehicle for which he has sought authorisation. At the time of writing vehicle identity discs are non-specific; that is they do not bear the registration number of the vehicle. They can therefore be interchanged between vehicles. Many operators have more vehicles in possession than they have authorisation to operate. Discs may be interchanged, but every vehicle used on the road must display a disc. Specific discs are under consideration, but at the moment Commissioners invoke Section 20 of the Public Passenger Vehicles Act 1981 which gives a Commissioner the power to request details of the registration numbers of vehicles in an operator's possession and that that information be kept up to date by the operator.

Traffic Examiners, who are part of the Vehicle Inspectorate, assist the Commissioner in ensuring that vehicles used on the road do display discs, that they are not overloaded, especially on continental trips and school contracts, and that the regulations governing drivers' hours and tachographs are observed. As with poor maintenance, operators who are convicted of drivers hours or tachograph offences will be brought before the Commissioner at an Inquiry on disciplinary grounds. The Commissioner for the West Midland and South Wales Traffic Areas considers it of utmost importance that any fines imposed by Magistrates for these offences are paid in full. He does not consider a fine to be a commercial transaction to be paid by way of instalments as and when funds are available and operators who take up this option cast doubt on the financial viability of their company.

Traffic Examiners also assist the Commissioner in monitoring local registered bus services. The Commissioner's role is to ensure that services run to the timetable and route lodged with him and that there is no anti competitive conduct by operators. The Commissioner is not empowered to comment on the suitability of routes or timetables or any other traffic related problems, these are matters for the local traffic authority.

However, under Section 7 of the Transport Act 1985 a local traffic authority may, in relation to a particular traffic problem, apply to the Commissioner for a traffic regulation condition which, if imposed, must be met by all holders of PSV licences operating bus services in the area to which the condition applies.

A traffic regulation condition may determine the route of services, the stopping places for services, when vehicles may stop at the stopping places and for how long, and any other matters as the Commissioner may feel to be appropriate. However, before the Commissioner imposes any such condition he must be satisfied, after considering the traffic in the area in question, that the conditions are required in order to prevent danger to road users or reduce severe congestion, and he must have regard to the interests of those who have registered services operating in the area, those who are or are likely to be users of such services and persons who are elderly or disabled.

The carrying of alcohol on coaches to certain sporting events is an offence under the Sporting Events (Control of Alocohol) Act 1985. Under this Act it is an offence for a person knowingly to cause or permit intoxicating liquors to be carried on a Public Service Vehicle (a) if he is the operator or the servant or agent of the operator or (b) if the vehicle is a hired vehicle and he is the person to whom it is hired, or the servant or agent of that person and the vehicle is being used to convey passengers to a designated sporting event. The police have found that the arrival of supporters at grounds hours in advance of kick-off time creates many problems and they have specifically asked for co-operation on the following:-

1. Coaches should arrive at the venue **no earlier than two hours before** the scheduled start but no later than one hour before the scheduled start.

2. Coaches attending designated sporting events should not stop within 10 miles of the final destination.

3. Coaches should not stop at any place where intoxicating liquor is available.

4. Departure from the venue should be within one hour of the end of the event.

The Commissioner expects operators carrying supporters to sporting events, especially football matches, to co-operate fully with the police.

Under Section 16(3) of the Public Passenger Vehicles Act 1981 the Traffic Commissioner has powers to place conditions on an operator's licence relating to journeys to and from designated sporting events. However, at this time the Commissioners do not use this power but the police have been asked to notify him of any incident where problems have occurred: he will then consider whether conditions should be imposed.

The Commissioner also, on behalf of the Secretary of State, regulates conduct of drivers and he has power to revoke or suspend a passenger carrying vehicle entitlement. All new entrants to the industry with four or more penalty points on their ordinary driving licence are warned as to their future conduct and in West Midlands and South Wales are personally seen. All drivers who are disqualified from holding an ordinary driving licence by a Magistrates' or Crown Court automatically have their vocational entitlement revoked and at the expiration of the period of disqualification can then re-apply for that entitlement. Such applications are determined by the Commissioner on behalf of the Secretary of State and the return is not necessarily immediate, nor is it automatic.

The role of the Commissioner is consequently an awesome one. He is responsible for the safety of each and every passenger carrying vehicle licensed in his Area. Whilst the Vehicle Inspectorate enforcement officers are available to investigate complaints and inspect vehicles, resources are not such that a check can possibly be kept on each and every operator and each and every vehicle. Thus the industry must be, by virtue of the circumstances, be self policing. The Bus and Coach Council plays an important part and with its officers leading by example higher standards can be attained and maintained.

The role of the Traffic Commissioner can be summed up in the single word 'Safety'. Safety is paramount.

March 1992

© JOHN MERVYN PUGH 1992

At times like these why pay more! MCW Metropolitan

All metal 45 seat luxury coach
£3,995

Inclusive of: Forced air ventilation. Ducted fresh air heating. Extra luggage locker. Chapman driver's seat. Windscreen washers. Mud flap.

Plus 2 years guarantee.
Try and beat that

mcw Metropolitan-Cammell-Weymann Limited, Elmdon Works, Marston Green, Birmingham

AP 303

BELFAST CORPORATION TRANSPORT. Depicted at Balmoral terminus in South Belfast in the mid-1960s are Harkness bodied Daimler CVG6 397 (OZ 6651) and Guy Arab III 302 (MZ 7400), the latter was used to provide a connecting service to Musgrove Park Hospital, as double-deckers could not pass under the nearby railway bridge.

BIRMINGHAM TO BELFAST

(and other places)

There is an unhappy tendency for Northern Ireland to be regarded as a foreign country whereas in reality it is more 'British' than the Isle of Man or the Channel Islands. It is logical, therefore, for the transport companies to prefer to buy British and indeed they showed more loyalty to home (i.e. UK) built vehicles than many mainland operators.

The first motor vehicles used in Northern Ireland were owned by the Midland Railway (Northern Counties Committee) as long ago as 1905. Another railway, the Great Northern (Ireland) ran extensive services from 1927 when these companies became legally entitled to operate buses. In the meantime Belfast Corporation had purchased a few vehicles but in the main relied on their tramway system. In 1935 a statutory body, the Northern Ireland Road Transport Board was given powers to acquire all the PSV services in the province (except Belfast) which they did to such an extent that by early 1936 they had taken over 687 buses belonging to 27 different companies.

In 1948 the N.I.R.T.B. was itself put under the umbrella of the Ulster Transport Authority whose brief included railways and commercial waterways giving in effect, a strengthened Northern Ireland version of the British Transport Commission (which body included British Road Services, British Railways and British Waterways) with absolute monopoly powers. Further changes and attempts at de-monopolization have taken place - the main one in 1966/7 when Ulsterbus took control of most UTA services and gave a new look to PSV services in Northern Ireland.

Many of the vehicles used by U.T.A. and its predecessors came from the Midlands and it is a pleasant thought that, say, a chassis from Wolverhampton could be transhipped, war or no war, bodied and then take men and women to the aeroplane factories and shipyards of Belfast - truly the sinews of war and peace.

BELFAST CORPORATION TRANSPORT. Sunbeam F4A/BTH Trolleybus 246 (2206 OI) with Harkness 68 seat bodywork. New in 1958 and withdrawn in May 1968, on the closure of the trolleybus system, after which it went for preservation to the British Trolleybus Museum.

CITYBUS. Marshall bodied 49 seat Daimler Roadline 1299 (7901 YZ previously registered KVT 173E) seen here working a Service 22 duty - Parliament Buildings - City Centre, is city bound despite shown destination. Originally in the fleet of Coastal Bus Services, Portrush, which was taken over by Ulsterbus in April 1974, it was later transferred to Citybus and was finally withdrawn in 1976. It was unique in being the only bus in the Citybus fleet with a Co. Londonderry registration.

One of the few independents is SURELINE COACHES, LURGAN. Mulliner bodied Guy Warrior (UWW 769), originally new to Morgan, Armthorpe, in 1957 - it was acquired in 1966 and withdrawn in 1970 when it was sold to a Gaelic Football Club in Coalisland.

This photograph is quite fascinating for UZ 7416 (L.416) is a Leyland Tiger Cub PSUCI/5T with an Ulster Transport Authority 41 seater coach body and has no obvious Midlands connection, but under the skin are frames from Metal Sections, Oldbury, who provided (and provide) many, many, thousands of Do-it-Yourself body kits to places as far apart as India and Africa. A few were used in the United Kingdom and this elegant machine, new in 1957, is a good advertisement for their products. The coach's condition says much for the quality control of the operators for maintaining and keeping clean a two-tone livery of Eau-de-nil and Cream required continuous attention by sympathetic employees.

ONE PERSON OPERATED - IN 1907.

This photograph is probably one of the most fascinating in this book, if only because of its age.

R.W. Kidner, the doyen of transport historians views it thus: "I suppose the photo shows an early attempt at a one-man-operated double-decker; it seems to have 20 on top and 12 inside and looks quite workable apart from what must be a rather narrow entrance between the stairs and the dash. The chassis is I think a Panhard of about 1907. I have seen no mention of such a bus in service; here it is not plated so presumably a trial run. As so often in these cases, the driver looks like the managing director. I guess the three men on top at the offside front corner are the Councillors or tramways officials, and the rest to make up numbers". Brakes looks as though they are of the so-called 'rim' pattern, directly operating on the wheels. The notice to the right of the door reads "Please pay fare" which backs up Roger Kidner's supposition. The Midlands connection lies in its having been photographed in Nottingham and part of what appears to be a tram is visible through the windows. However, the insignia on the side is that of the British Electric Traction Group, a magnet and a wheel.

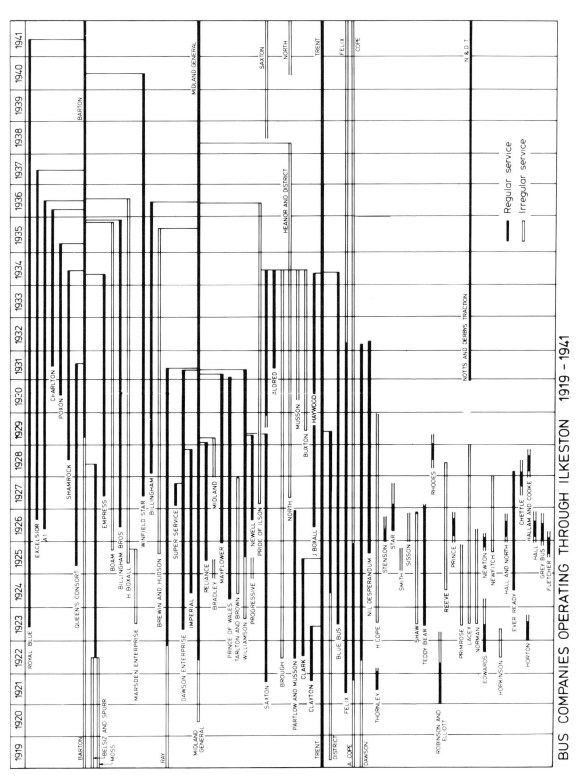

BUS COMPANIES OPERATING THROUGH ILKESTON 1919 - 1941

This magnificent table is reproduced from David Harrison's Primrose & Poppy, published by Paddock Publications in 1983 and is used by kind permission of the author and publisher. The plethora of operators in the 'free for all' days before licencing bit deep into the ranks of the independants is very obvious. The sad part is that modern deregulation, designed to re-open routes to all-comers has, instead, destroyed most of the survivors from those pioneering days.

[It will be noted that Brewin and Wilson, extant 1925-1935 share the same name as our publisher and any further information on this operator would be appreciated].

EAST MIDLAND & NORTH WESTERN

Visit Monsal Head
Four return journeys daily.

Service 58.—CHESTERFIELD—BASLOW—BUXTON.
Commencing on Saturday, August 4th, 1934, the above Service will be operated via Great Longstone and Monsal Head.

AMENDED TIME-TABLE DAILY.

	am	pm	pm	pm		am	pm	pm	pm
Chesterfield (Beetwell Street) *dep*	10 0	2 0	6 0	8 40	Buxton (Market Place) *dep*	9 25	1 45	5 25	8 45
Baslow (Devonshire Arms) *pass abt*	1027	2 27	6 27	9 5	Ashford Pump *pass abt*	9 59	2 19	5 59	9 19
Longstone (MONSAL HEAD) ,,	1044	2 44	6 44	9 22	Longstone (MONSAL HEAD) ,,	10 7	2 27	6 7	9 27
Ashford Pump .. ,,	1052	2 52	6 52	9 30	Baslow (Devonshire Arms) .. ,,	1024	2 44	6 24	9 44
Buxton (Market Place) *arr*	1126	3 26	7 26	10 4	Chesterfield (Beetwell Street) .. *arr*	1051	3 11	6 51	1011

NOR'WEST GOES EAST

If in the 1920s and 1930s a day out was sought the North Western Road Car Company's services into Derbyshire were to be recommended.

The company also had a penchant for slow but steady transport of somewhat spartan nature. Four of their prized specimens are shown; all Tillings-Stevens.

The body of the first is odd with its canvas roof, but it was one of a class of 30 delivered in 1924/5. The whole thing has a flimsy look about it, but a petrol engine which drove an electric motor gave the ultimate in gearless and jerkless transmission. Originally solid tyred, pneumatics were soon fitted but the class were withdrawn by 1932.

The next two vehicles are manifestations of the famous B10 class, of which over 2,000 were assembled. Most were powered by a 5.12 litre 4-cylinder petrol engine developing, on a good day, 63.5 bhp giving, for a 5 ton vehicle, the quite respectable consumption of 8-9 mpg. Many were later rebodied and some lasted until just postwar. 473 (DB 9373) was one of the Tillings built in 1930, rebodied by the Eastern Counties Omnibus Company (later Eastern Coach Works) with a 31 seat body in 1935. In this guise she lasted until 1946 and even in her last days would drift into Buxton, albeit slowly and with a plume of all-too-visible blue smoke.

No. 310 (DB 5210) is a pure 'bus' built in 1928 with little to soften either the exterior or the 36 seat interior. 557 (DB 9457) was a logical evolution. The outline has softened, the slight rake to the windscreen removing some of the boxiness and the destination boards would act as advertisements, albeit at the expense of considerable noise and drumming. Seating is more in keeping with an excursion vehicle and unlike 310, 557 has real headlights. The horn continues to protrude!

DISTRIBUTION OF P.S.V. OWNERSHIP, MARCH 31, 1936

WEST MIDLAND

	Vehicles
484 operators of up to 20 vehicles	1,266
11 operators of over 20 vehicles	2,197
495 operators	*3,463

* Of these Birmingham and Midland Motor Omnibus Co.Ltd., owns 902 and Birmingham Corporation 612.

Northern Scotland:-
384 operators of up to 24 vehicles
7 operators of over 25 vehicles

391 operators = 1,679 vehicles

Southern Scotland:-
252 operators of up to 25 vehicles
13 operators of over 26 vehicles

265 operators = 3,580 vehicles

Northern:-
331 operators of up to 19 vehicles =	1,058
15 operators of over 20 vehicles =	1,803
346 operators =	2,861

Yorkshire:-
442 operators of up to 19 vehicles =	1,169
25 operators of over 20 =	3,000
467 operators =	4,169

South Eastern:-
475 operators of up to 19 vehicles =	1,220
24 operators of over 20 vehicles =	3,299
499 operators =	4,519

North Western:-
4 larger companies =	2,255 vehicles
33 municipal undertakings =	2,325 vehicles
509 individual operators =	1,546 vehicles
546 operators =	6,126 vehicles

South Wales:-
399 operators of up to 16 vehicles
17 operators of up to 20 vehicles

416 operators

Western:-
649 operators of up to 19 vehicles	
15 operators of over 20 vehicles	
664 operators =	3,804 vehicles

Eastern:-
386 operators of up to 19 vehicles	1,037
8 operators of over 20 vehicles	1,087
394 operators =	2,124 vehicles

Metropolitan:-
394 operators (all independent) of up to 19 vehicles
14 operators (all independent) of over 20 vehicles

408 operators (excluding London Transport)

THE RURAL PROBLEM

Unfortunately in the 1930s methods of collating information varied between the different Traffic Commissioners offices and then and now the boundaries of their areas do not entirely align with those of a 'Midlands' book. The West Midlands area is highlighted but the figures should be read in conjunction with neighbouring areas.

"Has the time come to call a halt in the process of monopolization in the road passenger-transport industry? Is the way being paved towards nationalization? Is the convenience of the whole British public being jeopardized by the growth of large inter-connected combines and statutory monopolies?" Not 1955, 1975 or even 1992, but words which appeared in **Commercial Motor** during 1937.

The arguments used bear an uncanny and uncomfortable resemblance to those of 1992 but PSV professionals were then worried on the one hand over the growth of London Transport-type combines and on the other the decline of small operators since the passing of the 1930 Transport Act. On a number of issues **Commercial Motor** explained the pros and cons of large operators. Statistics culled from these and other articles are interesting. In 1933 there were 5936 operators, of whom 5747 were 'tiddlers' with fleets of up to 24 vehicles each. Many, indeed, were owner-operators who were to fall foul of either the depression (as they have in the 1990s) or were unable to compete with undercutting of fares by the big concerns. By 1935 the number of operators had dwindled by 623 (10.5%) with all of this decline coming from the 'tiddlers', 625 (10.9%) of them having gone to the wall.

The loss of these small firms led to a decrease in the 'small operators' fleet strength of 1,257 but the bigger concerns rose by no less than 2361. To summarize: in 1933 5747 small operators owned 15,964 vehicles whereas in 1935 5313 small operators owned 14,707 vehicles but in 1933 179 large operators owned 29,369 vehicles whereas in 1935 178 large operators owned 31,730 vehicles. The biggest argument against these large groupings was, and is, that the once intensive local bus service must continue to decline. We can illustrate this best by an imaginary example.

A small operator, which we may call Gladwin's Luxury Coaches started operations back in 1920 from the village Upside to the town Downside using a bus-o-lorry with its bus body being fitted for Market days and Sundays and a lorry body for carrying coal on the other days.

Eventually the firm progressed to a real bus-that-could-serve-as-a-coach and in between trips to the coast began to run daily alternative services from Upside to Downside direct or via a ten mile diversion through Middleside, a hamlet of 100 people or so. This route extension lost money on the actual bus service but provided odd coachloads for outings and Sunday School trips. Passenger loads remained static until the early 1930s but as the slump bit declining traffic forced Gladwins to realize these bus routes were no longer profitable by any standard. When the Pobble Bus Company offered some financial incentive to take them over it has to be admitted that, like most of his colleagues, the Managing Director of Gladwins, by now 'young' Mr. G, seized it with alacrity, particularly as the coach operations were to remain in his hands.

Acquiescing to this and other takeovers, the East Midlands Commissioners were scathing: "Complaints are frequently received that when the services of small operators are taken over by the large operators the public find that the facilities they previously enjoyed have been considerably reduced". Quite so, for the first action of the Pobble Bus Company, who only wanted the main road service, was to eliminate the diversion via Middleside.

Sidney, a farm worker was accustomed to walking to the bus stop in Middleside and catching the 6.30 a.m. bus to Upside Farm. When the service stopped he had either to cycle eleven miles, stop working or cycle four miles to the main road and catch the 6.15 a.m. to Upside Farm. But there was nowhere to leave his bike and inevitably when he couldn't cycle to work in mid-winter he joined the unemployed. Referring again to the growth of large firms and the complaints flooding in the Commissioners stated in 1937: "We have found it necessary resolutely to check the attempts to withdraw facilities from the travelling public, which have resulted directly from such absorptions". Too late though for Sidney and thousands like him.

But who was to blame? Today we can glibly blame motorcar traffic as yet another Midlands route is lost, but really the malaise is too deep seated for such facile arguments.

In 1937 the question was asked "Will the time come when the road passenger-transport industry is controlled by 100 or so inter-linked private companies and a similar number of municipalities or [by] fewer but larger transport boards?" In the West Midlands area we saw how Birmingham Corporation tightened their grip on services, with only Midland Red as a real partner. There were, and are, many peripheral services but few operators regularly worked buses into Birmingham. Then, after nationalization came the WMPTE and the Midland 'Polo' Red for loosing their inner Birmingham routes to the WMPTE they became like a mint with a hole in the middle. Now with de-regulation, millions of pounds are paid out by Centro (the local traffic authority) - to prop up unviable routes, while firms fight for the minty bits that can make a profit.

Municipal services throughout the country are almost extinct but the fears of 1937 are very, very real in the 1990s, for leaner, fiercely competitive companies are liable to look at loss making rural or urban routes with only a cold, cold accountant's eye and Councils cannot subsidize all these routes for ever.

APPLICATIONS FOR PSV OPERATORS' LICENCES - DURING THE YEAR ENDING 31 MARCH 1991

TRAFFIC AREA	Received during year	Carried over from previous year	Total	Issued	Refused	Withdrawn	Awaiting decision	Total number of licences in issue
NORTH EASTERN	243	80	323	208	7	14	94	953
NORTH WESTERN	322	42	364	310	5	11	38	945
WEST MIDLAND	**141**	**27**	**168**	**122**	**1**	**23**	**22**	**457**
EASTERN	255	70	325	206	1	34	84	905
SOUTH WALES	121	13	134	101	2	14	17	442
WESTERN	211	60	271	193	0	15	63	678
SOUTH EASTERN	171	46	217	161	1	17	38	598
METROPOLITAN	187	86	273	194	1	1	77	804
SCOTTISH	232	82	314	198	3	7	106	865
TOTAL	1883	506	2389	1693	21	136	539	6647

ALLCHIN'S
LUXURY COACHWAYS

Midland Service & Time Table

From OXFORD
DAILY, JUNE TO OCTOBER, 1933.

(†After OCTOBER 31st—Down, Saturday & Monday. Up, Friday & Saturday)

FARE S.	FARE R.	READ DOWN		Departure from Cattle Market... Daily (as above).		READ UP	
		A.M.	† P.M.			†	
—	—	9 0	3 30	OXFORD	Gloucester Green	12 50	7 20
2/-	3/6	9 55	4 30	BRACKLEY	H. W. Plank, Market Square	11 50	6 25
4/3	7/3	10 50	5 30	NORTHAMPTON‡		10 50	5 30
—	—	9 0	3 30	OXFORD	Gloucester Green	12 50	7 20
6/6	11/-	12 0	6 20	MARKET HARBORO'	The Square	9 57	4 37
7/-	12/-	12 18	6 35	KIBWORTH	The Stores, Leicester Rd.	9 42	4 20
7/6	12/-	12 39	6 55	OADBY	Post Office	9 21	4 1
7/6	12/-	12 45	7 0	LEICESTER	Car Park, London Road	9 15	3 56
7/9	13/6	1 4	7 20	MOUNT SORREL	Allen's Garage	8 56	3 37
8/6	14/-	1 15	7 32	LOUGHBOROUGH	Market Place	8 45	3 26
9/6	14/-	2 0	8 15	NOTTINGHAM	Richardson's Car Pk., Derby Rd.	8 0	2 45
—	—	—	3 30	OXFORD	Gloucester Green	12 50	—
9/6	14/-	—	7 50	KEGWORTH	Post Office	8 25	—
9/6	14/-	—	8 5	LONG EATON	Market Place	8 10	—
9/6	14/-	—	8 15	SPONDON	Corner Shop	8 0	—
9/6	14/-	—	8 30	DERBY	Sanderson & Holmes, London Rd.	7 45	—
—	—	9 0	3 30	OXFORD	Gloucester Green	12 50	7 20
5/6	8/3	11 37	6 0	WELLINGBOROUGH	Lloyds Bank	10 12	4 58
6/-	10/6	11 58	6 22	KETTERING	Headland's Garage	9 52	4 38
7/-	13/6	12 23	6 49	THRAPSTON	Bridge Street	9 27	4 13
8/6	14/-	12 43	7 11	OUNDLE	The Square	9 7	3 52
8/9	14/-	1 20	7 50	PETERBOROUGH	27 Long Causeway	8 30	3 15
—	—	—	3 30	OXFORD	Gloucester Green	12 50	—
6/3	10/9	—	6 30	RUGBY	40 Church Street	9 45	—
7/-	11/6	—	7 5	COVENTRY	Pool Meadow Car Park	9 10	—
8/9	14/6	—	8 0	BIRMINGHAM	Dale End Garage	8 15	—
—	—	9 0	—	OXFORD	Gloucester Green	—	7 20
2/-	3/6	9 55	—	BRACKLEY	Plank, Market Square	—	6 30
2/9	4/9	10 20	—	TOWCESTER	Market Square	—	5 55
4/3	7/3	10 50	—	NORTHAMPTON‡		—	5 0
6/3	10/9	11 50	—	BEDFORD	St. Peter's Corner	—	4 0
8/9	14/3	12 25	—	ST. NEOTS	Market Place	—	3 5
9/3	15/9	1 14	—	CAMBRIDGE	Maids Causeway	—	2 35
9/9	16/9	1 51 / 2 30	—	NEWMARKET	Crown Hotel	—	2 0 / 1 26
11/3	20/3	3 25	—	THETFORD	Lambert, Norwich Road	—	12 30
13/3	23/3	4 5	—	ATTLEBORO	Church Street	—	11 52
14/3	23/9	4 22	—	WYMONDHAM	Market Place	—	11 35
14/3	24/9	4 46	—	NORWICH	Bell Avenue Car Park	—	11 10
16/3	27/3	5 42	—	GT. YARMOUTH	St. Nicholas Church	—	10 15
16/9	27/9	5 50	—	GORLESTON	Railway Approach	—	9 52
17/9	28/9	6 12	—	LOWESTOFT	Marine Car Park	—	9 30
						A.M.	P.M.

‡Departure from Northampton. Wednesdays and Saturdays—Campbell Square.

Other Days—Market Square.

THE MARKET BUS - LOST BUT NOT FORGOTTEN

The market bus was at its ascendancy from the late 1920s, a period the following extract refers to, until the late 1950s when private motoring and the replacement of the cinema with television led to its decline in all bar a few remote areas.

During the writing of this book two drivers belonging to a private firm emulated their predecessors by travelling on the country market-day services in Worcestershire and Herefordshire they had known in the 1960s. Apart from schoolchildren the maximum load never exceeded ten, vehicles were in the main the ubiquitous Leyland National, unkempt and inevitably pumping out smoke. Out of some 20 trips only twice did the driver vacate his seat, once for a woman with an artificial leg, and once for a blind man. For the rest, passengers young and elderly alike were left to hump their shopping bags, prams and trolleys on board as best they could. The National was designed as a fast flow city bus; and although the economics of 'cascading' are well understood, the old Duple Midland bodied Bedford or Commer, or the famous Midland Red bus was so much easier to board and alight that there is no comparison.

The following quotation is from Mona M. Morgan; **Growing up in Kilvert Country**, published by the Gomer Press, Llandysul, Dyfed; 1990. It is reproduced by permission of author and publisher to whom we extend our thanks.

It was the market-bus that brought most benefit and pleasure, especially to small-holders and cottagers who had previously been obliged to walk to town on market days. Those with traps used them less and less until they were finally relegated to some dusty shed, there, with shafts aloft, to end their days.

The bus ride itself was no match for an open-air trap ride on a fine day. It was the company that everyone so enjoyed. Market journeys were jolly, social occasions, rather like day-trips, with friendly, animated country folk laughing and joking and calling to each other across the bus. Everyone was welcomed aboard with nods and smiles of greeting and a quip or two from the wits. The drivers were cheerful, willing chaps, ready to carry messages or undertake shopping orders for all and sundry.

Farmers' wives came laden with baskets of poultry, butter and eggs. One particularly inquisitive old dame always wanted to know, of those around her, 'What 'ave you got taking today?' On the return journey she quizzed people on what they had made on their produce. Folks tried to ignore her but she persisted till she got an answer. But when she grilled her neighbour on how much he made on his turkeys at Christmas he said he didn't know. 'Aye but you *do* knah,' she insisted. 'Well I sha' tell ya then,' was his brusque reply.

On the return journey, baskets laden with provisions were stacked aboard, everyone lending a hand. News gathered during the day was shared and lively animated talk echoed around. The eyes of the market-piert were a touch brighter than normal, their loosened tongues additionally entertaining. All passengers alighted to goodbyes, good wishes and jocular advice on how to conduct themselves in the coming days.

In later years city folk on holiday in the district were fascinated by the market journeys. They were mystified when, for no apparent reason, the driver pulled up at a field gate, until they heard him remark, after several minutes, 'No sign o' Mrs Jones yet, but she'll be 'ere in a minute, sure to.' As she was spotted running and panting with her heavy baskets some might jocularly shout, 'Come on, Mrs Jones, pierten-up or it'll be time to come back afore we gets there.' Climbing aboard she might say, "Ow long you bin waitin', Jack? Our clock was stood and I didna know no aim what the time was.' Such a remark would invite a bit of leg-pulling from the men on board and some spirited rejoinders from Mrs Jones. There would probably be further waits at gates or road-junctions, to make sure no one was left behind. Perhaps some one would be waiting to hand the driver a shopping-list, with the request, 'Please to bring these few things for me, Jack,' sure of a willing response. To city folk it all seemed so friendly and matey and, though unfamiliar with the local parlance, they loved the fun and repartee, and many were forced to revise their previous concept of the countryside as a dull and lonely place in which to live.

Charlton-on-Otmoor services must have one of the most enviable locations in the Midlands. Two views of their garage.

Rarely did a "Market Bus" make itself as known as well via its blinds as this down-graded coach.

Bartons specialized in 'horses for courses', not disdaining lightweight chassis for bus work. This is a Reo and advertises "Skegness twice daily" in the window.

34

A meeting place while passengers are waiting for the bus; or, today, while waiting for the driver to unlock the door.

Market-day buses in their lair. Worcester Garage, Midland Red, in the early 1950s.

Banbury one market day in 1991 and a Duple bodied Dennis discharges a capacity load.

Ubiquitious National and departing family. As it was in school term time, this must have been a "teachers training day"!

Destination information can be somewhat sparse. PEX is a long way from her one-time home in Great Yarmouth.

Down-graded coach again, with even the vinyl applied skew-whiff!

MUNICIPAL TROLLEY BUS RECORDS

(from Transport World January 17 1935)

SYSTEM	Year Ending	Total Capital Outlay To date	Total Operating Income	Total Operating Expenses	Percentage of expenses to Receipts	Gross Profit	Net Surplus	Total Bus Miles Run	Average Per Bus Miles Total	Revenue Miles Traffic	Average Fare Per mile	Average Fare Paid Per Passenger	Average Distance of 1d Fare Miles
		£	£	£		£	£		d.	d.	d.	d.	
BIRMINGHAM		184,530	42,524	31,438	73.98	11,086	A	607,954	16.787	16.787	0.72	1.189	1.33
CHESTERFIELD		50,903	32,968	21,084	60.90	11,885	3,015	588,536	13.44	13.33	0.98	1.41	0.78
DERBY	MAR	A	82,466	57,436	69.65	25,030	10,828	1,178,641	16.79	16.789	0.98	1.476	0.79
NOTTINGHAM	31	246,178	122,910	91,551	74.49	31,359	7,547	1,787,940	16.498	16.403	0.81	1.239	1.02
WALSALL	1934	60,457	27,755	16,367	58.92	11,388	6,565	418,459	15.931	15.806	1.098	1.607	0.84
WOLVERHAMPTON		575,284	235,194	151,372	64.36	83,822	AB	3,729,699	15.134	15.015	1.179	1.645	0.90

Note 1d = 2.4p. But figures are comparative	OPERATING EXPENSES PER BUS MILE							TROLLEY BUSES							
	Traffic d.	General d.	Licences d.	Repairs & Maintenance d.	Power d.	Miscellaneous d.	Total d.	Length of Route Miles	Average miles per day per bus	Average speed per hour	In Use Daily	In Stock SD	In Stock DD	Average Seating Capacity	Total Passengers Carried	Passengers per Bus Mile
BIRMINGHAM	6.372	1.204	0.940	1.602	1.944	0.349	12.411	7.08	86	8.73	54	—	66	56	8,582 408	14 12
CHESTERFIELD	3.96	1.34	0.43	1.32	1.49	0.06	8.60	5.5	100	12.00	12	14	2	34	5,552 016	9.43
DERBY	5.172	0.972	0.900	2.236	2.031	0.383	11.694	17.10	149	9.22	28	—	56	56	13 293,053	11.28
NOTTINGHAM	5.694	2.035	0.663	2.075	1.822	—	12.289	14.65	113	8.66	43	—	50	60	23,148,275	12 95
WALSALL	4.661	1.05	0.683	1.283	1.710	—	9.387	9.929	108	10.98	16	—	19	60	4,115,355	9.83
WOLVERHAMPTON	4.619	1.310	0.572	2.102	1.137	—	9.74	46	125	9.38	82	32	73	53	34 215,694	9.17

NOTE: A = NO SEPARATE ACCOUNTS ARE PUBLISHED FOR EACH DEPARTMENT
B = NET SURPLUS ON OMNIBUSES AND TROLLEY BUSES £15,581
SD = SINGLE DECK
DD = DOUBLE DECK

Nottingham 481, seen here, was one of a small batch of Karrier WS, 479-82, delivered in 1948 with Roe bodywork. The entire batch was withdrawn in 1965, the same year that the 47 route to Ransom Road was converted to motorbus operation on 9 October.

Much of the centre of Nottingham has been altered drastically since the 1960s; the area around the former Great Central Railway Station at Nottingham Victoria is no exception. Roe bodied Karrier W 460 of 1945 is caught passing through this area on route 40, which served the suburbs south-west of the city, forming the southernmost point of the system. The route converted to motorbus operation on 9 October 1965.

WHO'S A LITTLE SUNBEAM THEN?
ANDY SIMPSON

From 1922 to 1970, travellers through the Midlands' larger industrial towns and cities stood a great chance of coming across blue, green or green and yellow trolleybuses in crowded urban centres, leafy suburbs and on lengthy interurban routes. This article attempts to give a brief overview of the relevant systems and, in particular, their connection with the Sunbeam company.

The southernmost Midlands Trolleybus system was that of the second city, Birmingham. Birmingham's dark blue and primrose trolleybuses started a trend, as the first route to Nechells, operated from 27 November 1922, was the first tram to trolleybus conversion in the country. The only other Birmingham route, the 5¼ mile run to Yardley, opened in January 1934. That same month Birmingham's only Sunbeam trolleybus - No. 67 (OC 6567) was delivered on loan from the makers. She was a MS2 six wheeler with Metro-Cammell bodywork. Although running as a company-owned demonstrator (of which Birmingham had several at various times) she was operated in full Birmingham livery, albeit for only a brief period until retired to Sunbeam who in September that same year sold her to Wolverhampton as their No. 222 - the only secondhand (as opposed to loaned) vehicle to work in Wolverhampton.

The Birmingham system itself, despite wartime extensions, closed in June 1951, when the mainly Leyland fleet went for scrap.

Although a journey from Birmingham to its northern neighbour Wolverhampton was once possible by tram (involving several operators and changes of vehicle) the two centres were never connected by trolleybus. Wolverhampton, standing on the northern edge of the Black Country but always holding itself a little aloof, was, however, an early starter in the trolleybus stakes. Inspired by Birmingham's conversion of the Nechells route, Wolverhampton converted its 2¼ mile (3.6 km) urban route to Wednesfield to trolleybus operation on 29 October 1923 - the second tram-trolleybus in the country. The 3 mile (4.8 km) route to Fordhouses followed next (March 1925) and the 5 mile (8 km) run to Dudley in 1927. In 1929 Wolverhampton briefly held the title of the world's largest trolleybus operator. Expansion was almost complete by 1935, with completion of routes serving the growing suburbs. At it's peak, Wolverhampton operated 45½ route miles (73 km) of trolleybus services, with routes radiating out along eleven main roads to 15 town termini. Despite extensive post-war fleet enlargement, an ageing infrastructure brought about a staged abandonment. The decision to abandon was taken in 1961; and the replacement programme commenced 1962. Main abandonment of routes was from June 1963 (lines to Penn), completion in March 1967 (Dudley route).

Wolverhampton purchased a total of 354 trolleybuses over 27 years; 173 of these were Sunbeams from 15 repeat orders, 153 Guy/Sunbeam vehicles were in stock at the end of 1961.

The first ever Sunbeam Trolleybus, built 1931, entered service in 1933 as No. 95, after use as a demonstrator. The earliest Wolverhampton vehicles were Tilling-Stevens single decks; from 1926 the 32 deliveries were exclusively by Guy BTX double deck vehicles; from 1932 to 1950 the local manufacturers, Guy and Sunbeam shared all orders.

1934 saw delivery of the prototype Sunbeam MF1 4-wheeler (27' body), No. 206, one of a batch of 4 single deck vehicles; further Sunbeam MF2 single decks, 231-3 followed in 1936. From 1938-42 Wolverhampton took delivery of 38 MF2 double deckers, of which 20 were sold 1950-52 for further service in Belfast (11) and Southend (9). Wartime Sunbeam W utility deliveries were 296-9 (1943); 400-1, 402-7 (1944) and 408-18 (1945).

Postwar, the ageing fleet of 1930s vehicles was renewed up to 1950 by large numbers of Guy BT and Sunbeam vehicles. The fleet was further updated by rebodying some vehicles in 1962 (Park-Royal) and 1958-62 (Roe) - a total of 54 vehicles, including 37 of the Sunbeam Ws delivered 1943-8, given Roe bodies. Survivors include 1946 Sunbeam W 433 (rebodied Roe 1959 - runs regularly at the Black Country Museum) 1949 F4 616 and 1950 Guy 654.

A few miles east of Wolverhampton, and linked to it by a well-known joint trolleybus route, lay the leather-working town of Walsall, with a trolleybus system that had a flavour all of its own. Powers to operate trolleybuses were acquired in 1914 and consolidated in 1925. Operations began on the joint route to Wolverhampton in November 1931 (succeeding short workings to Willenhall from that July). In 1933, 15 Sunbeam MS2s were acquired for conversion of the Bloxwich tram route (Walsall's last). Six more MS2s arrived in 1938. Like its neighbour Wolverhampton, Walsall borrowed 1934 Sunbeam MS2s from Bournemouth to brighten the wartime darkness, and managed to acquire its first four wheelers - the inevitable Sunbeam Ws - in 1943 (two) and a further six in 1945, still all operating over the original two routes. Ten Brush bodied F4s, 334-43, followed in 1951, 334 being exhibited at the 1950 Commercial Motor Show - this batch formed the most silent and smooth running buses in Walsall. In 1960 342 was lengthened using a Bristol Lodekka diesel bus chassis. This was yet another innovation by the Walsall manager, Mr. R. Edgeley Cox, who took over the reins in June 1952. He had also used S7 850 to experiment with a Willowbrook passenger flow body, having stairs/exit in the centre of the body. Slow loading caused a rebuilding. Crews called her the "Queen Mary".

In November 1954, Walsall caused a sensation with the introduction of the first of its new 30 type F4a, 851, with 70 seat Willowbrook 'Goldfish Bowl' bodies. The crews called these two batches of vehicles, delivered 1954-1956, "Liners". Other names abounded - "Big Brothers" (the pre-war six-wheelers); "Sardine Cans" (ex-Hastings Ws) and "Rattle Cars" (utility bodied vehicles). Under Cox's leadership second-hand vehicles were acquired and new routes opened, right up to September 1963 (Cavendish Road) - the route mileage more than doubled 1955-1963, giving a maximum of 18.86 route miles.

Innovations, and plans for extension, continued throughout the 1960s. In 1969, "Liner" 866 was rebuilt for front entrance format; 5 additional routes were planned in 1962. Bournemouth MF2B 300 was borrowed in January 1969, with the hope of acquiring many, if not all, of the class, since the Bournemouth system was about to close. No. 300 was heavily modified - her Weymann bodywork had its rear entrance and staircase removed, an emergency exit provided on the offside, and both saloons extended to the rear of the vehicle, making it capable of OMO. However, it never entered service. All plans for extension were ordered by Walsall but cancelled by the newly formed West Midlands PTE on 1 October 1969. The PTE had little time for Walsall's 46 trolleys amongst its otherwise all motorbus fleet. No. 300 headed north to a Barnsley breakers yard, soon followed by all but the 7 of the Walsall fleet which survive in preservation.

Most second-hand vehicles came off the road on 13 February 1970; normal service ended 2 October 1970. The following night Edgley Cox himself drove the last trolley, 872, into Birchills depot, closing the last trolleybus system in the Midlands.

Three of the 'Liners' - 862/4/72 survive - appropriate since they were Britain's last 30' long double-deckers, Edgley Cox negotiating with the Ministry of Transport for the necessary permission to operate. 862 operates most weekends at the Black Country Museum, Dudley, on what is currently Britain's longest trolleybus route. The steep gradient up to the depot gives her - and Wolverhampton 433 - the chance to show their paces.

Moving further north, we find the City of Derby, whose first trolleybus route, running 2½ miles (4 km) along Nottingham Road, opened on 9 January 1932. Further extensions came 1932-34 as part of a programme of tramway replacement. New routes came also in 1943 with a short extension along Sinbin Lane, and postwar during 1947-1958, giving a total route mileage of approximately 28.

In addition to the typical urban centre and suburbs the Kingscroft route offered a vivid contrast in scenery - the section to Wyndham Street included both open country and the famous railway works.

A total of 70 trolleybuses were purchased 1932-5 as tram replacements, all Guy BTX types. Utility Sunbeam Ws jointed the fleet in 1944. Utility 171 was the first

37

Sunbeam in the fleet, other than trials 'bus 100. All subsequent orders were for Sunbeams - a total of 74 Sunbeam chassis were delivered including 7 repeat orders. No. 100, a Dodson bodied Sunbeam MS2, ran from 1932-1951 and featured Lockheed brakes.

1944 Weyman bodied wooden slat seated Ws 171-2, of which the latter vehicle is still preserved, were followed in 1945 by 'relaxed utilities' 173-5, and in 1946 by Ws 176-85, with Park Royal bodies. In 1948/9 came the first 8' wide vehicles in the fleet, 186-215, Brush bodied Sunbeam F4s. These were followed in 1952-3 by F4s 216-235, the first Willowbrook bodied trolleybuses ordered from Brush but built by Willowbrook after their takeover of the body-building activities. Finally in 1960 came the 8 F4s, with Roe bodies, 236-43, delivered as replacements for 1938 Daimlers. These were the last additions to the trolleybus fleet, giving a fleet total by 1962 of 73 vehicles when conversion to motorbuses began. The process was completed on 9 September 1967, the crews marking the occasion by decorating 'their' vehicles. Wiring demolition began within two days, but five vehicles survive to remind us of this system. As happened elsewhere, the end was hastened by the introduction of traffic management and one-way systems. 26 vehicles survived to the bitter end.

Derby was linked to Nottingham by the lengthy BET owned Notts & Derby system whose fleet never included any Guy, Karrier or Sunbeam vehicles.

Nottingham itself in the early 1930s, boasted Britain's largest trolleybus fleet, traversing the typical red-brick lined streets of a Midlands industrial town. Its lack of scenic spots was partly compensated for by speedy operation. The towns narrow, hilly roads saw rush hour services of 2-3 minute intervals over long sections of route.

The story began in April 1927; traditional conversion of tramway routes saw the final trolleybus extension in 1935, a year before the last tram ran. The fleet included Ransomes, Simms & Jeffries, Karrier, Leyland, BUT and Sunbeam vehicles. In 1942 came five new 8' wide Sunbeams 447-51 - the first 8-footers in the fleet - originally ordered for Johannesburg but diverted due to the war. These were MF2 chassis. Karrier Ws followed in 1943-8, Nos. 442-5 and 452-82. Number 476 of 1945 was the first with automatic acceleration, and carried Roe bodywork. BUT vehicles made up most post-war deliveries, the last being in 1952. A total of 40 Sunbeam/Karrier trolleys went to Nottingham. The decision to abandon was taken in 1961, when there were 138 trolleys in the fleet and 300+ motorbuses. The first closures came in 1962, the bulk in 1965, in which year all remaining 2-axled trolleys including the last utilities were withdrawn on the dieselisation of route 43 Trent Bridge-Bulwell. Three-axle BUTs saw out the final months until closure on 30 June 1966, with a special closing ceremony the following day. The six preserved survivors include 1945 Karrier W 466 at Sandtoft, where the other survivors have also congregated.

© Andy Simpson 1992

One of the last batch of trolleybuses delivered to Birmingham No.85, a Metro-Cammell bodied Leyland TB7, could seat 54 passengers and was powered by an 80 h.p. G.E.C. motor. The notices on the right contain admonitions to drivers not to obstruct crossroads, and conductors to give clear hand signals (no trafficators being fitted).

Seen here in typically battered Walsall condition, original 1933 Sunbeam MS2 161 takes a breather in Park Street, opposite the railway station. With her Short built H32/28R body she served for 18 years, succumbing to old age in 1951. In this view she is, appropriately enough, bound for Bloxwich. She had been delivered as a replacement for the Bloxwich trams, Walsall's final tram route.

Walsall 'Goldfish bowl' Sunbeam F4A 871, is seen here on circular route 30 - Walsall (bus station) - Leamore-Bloxwich-Blakenhall-Walsall (Bus station) in t'.e late 1960s. Entering service on 1 October 1956, she was withdrawn 31 March 1970 and passed to Wombwell Diesels, Barnsley, for breaking the following November together with 18 of her sisters. Her distinctive Willowbrook bodywork was visually in a class of its own!

A rather unorthodox view of an all-metal trolleybus body is shown here at the Brush Coachworks, Loughborough, September 1935. The basic skeleton of the lower deck is complete and the truss plate, holding the structure together and giving it its strength, has been riveted up. The mechanics are from Leyland Motors and the body on completion will seat 60. Low voltage lighting will be utilized and the batteries will be so designed that they can (given maintenance) propel the 'bus in an emergency. Air and reostatic brakes ensured safety.

Seen here at the hub of the Wolverhampton system - Victoria Square, in the late 1930s are two classic Sunbeam six-wheelers. Leading the twosome is 222 - formerly Birmingham 67, and Wolverhampton's only second-hand trolleybus purchase. She is followed by 1936 Sunbeam MS2 No. 226, with a Park-Royal 58 seat body. This vehicle was withdrawn c.1948 and moved to a site near Post Office Road, Seisdon, for an afterlife as a caravan, finally being broken up c.1967-8.

A classic Wolverhampton combination is seen here in Stafford Street - 441, a 1947 Sunbeam W, rebodied by Roe, leads the line-up; she is southbound for the suburban terminus of Merry Hill, and lasted until February 1967, when she was cannibalised for spares during the last month of Wolverhampton's trolleybus operation, and passed to Hammell, Bloxwich, for breaking. She is followed by 463 - a 1948 Sunbeam F4 with Park-Royal body on route 12 to Finchfield, which closed together with route 13, in Novmeber 1963.

Seen here amongst the dingy delights of Pinfold Street, Darlaston, Terminus, Wolverhampton 633 - a 1949 Guy BT with a Park-Royal body - is waiting to run via Bilston back through Wolverhampton and to the Victorian suburb of Whitmore Pears. 633 was one of the Guy vehicles built by Sunbeam when absorbed by Guy Motors.

Wolverhampton's 455, a 1948 Sunbeam W, delivered with Park-Royal bodywork but later rebodied by Roe c.1962, is seen here at the Dudley terminus of the 5-mile long route to Wolverhampton which was Wolverhampton's last route, and the only trolleybus terminus in Worcestershire! Repainted in 1965, 455 ran on the last day - Sunday 5 March 1967 - and passed to Gammell, Walsall, for breaking by 1 April that year. She was identical to the preserved example, No. 433, now running at the Black Country Museum, Dudley, on what is currently Britain's longest trolleybus route, providing internal transport on the museum site.

Midlands trolleybus operators sold very few vehicles for further service with other operators, although in 1950 Wolverhampton sold its 1938 batch of Sunbeam MF2s, Nos. 264-269, to Southend. These carried Park-Royal H28/26R bodies, and Southend 148, ex Wolverhampton 268, is seen here in the depot yard at Southend (London Road) from where it and its sisters ran until closure of the system in October 1954. Other Wolverhampton MF2s went to Belfast in 1952 but did not last long. Another ex-Wolverhampton vehicle sneaks in at the rear.

DERBY TROLLEYBUSES

ARC 515 on route 60. Buses galore including a Trent Motor Traction Atlantean sneaking into view at the rear, as the now preserved 215 crosses the hub of the Derby system, the Market Place, en route to Osmaston Road, location of one of the system's three depots. No. 215, a 1949 Sunbeam F4 with Brush bodywork has run in preservation at Crich Tramway Museum and the Black Country Museum, Dudley.

ARC 506 on route 11. Love at first sight? Derby pedestrians inspect the photographer, and each other, whilst waiting for 1948/49 built, Brush bodied Sunbeam F4 206 to pass. This batch Nos. 186-215, were the first 8 foot wide vehicles in the fleet. Nos. 201-215 were also fitted with traction batteries for off-the-wires manoeuvres. No. 206 itself was withdrawn in 1966 and is seen outbound on the No. 11 route to Allestree Lane via Kedleston Road, having begun her journey at the L.M.R. station on London Road.

ARC 489 on route 43. Heading the line-up in the Cornmarket is No.189, a Sunbeam F4 of 1948 with Brush bodywork. She is heading for Drutheld Road (Kingscroft) terminus, the northernmost point of the system, and doubtless a familiar destination until her withdrawal in 1966.

RC 8879 on route 41. Seen here out in the suburbs on the 41 route, Market Place-Alvaston-Harvey Road (Mitre) is Park-Royal bodied Sunbeam W 179, delivered in 1946. The 176-185 batch re-introduced provision for route numbers, missing from previous batches of 'Ws'. No. 179 lasted until 1965.

40

SIX TOWNS TRANSPORT LTD

During the late 1930s one of the major talking points among Midland bus operators was the proposed scheme for a compulsory merger of operators within an area bounded by Stafford to the south, Market Drayton to the west, Congleton and Utoxeter; the whole being centred on Newcastle-under-Lyme, although it was expected administration would be based on Stoke-upon-Trent. A similar plan, drawn up in the wake of London Transport monopoly in London, was for another group - South-east Lancashire-North-east Cheshire - to have a rather lopsided area based on Manchester but extending to Bolton, Rochdale and Stockport. The acronym SELNEC was obvious and bus transport in this area was to be finally consolidated under this banner on 1st November 1969, over thirty years after its planning.

By contrast the Staffordshire grouping never really got off the ground. After many discussions by 1937 the outline of the plan had hardened and incorporated all operators within a dozen miles (20 km) of the Six Towns plus it was intended to seek general powers to run 'contract carriages' to places within a radius of ten miles (16 km) outside their planned boundary.

This transport concern which would, of course, have a total monopoly of bus services was to be administered by a Board consisting of 11 delegates from Stoke-on-Trent, 6 from Newcastle, one from the boroughs of Crewe, Stafford and Congleton, and one each from the Urban Councils of Alsager, Biddulph, Kidsgrove, Leek, Market Drayton, Stoke and Utoxeter. The Rural Districts represented were to be Cheadle, Leek, Newcastle, Stone and Utoxeter. Rather surprisingly the planners had no desire to operate coach services on any routes to destinations outside their perimeter.

They would have complete and total powers to decide timetables and fare structures and the sponsors (basically Stoke and Newcastle, who agreed to cover any losses) believed that with an injection of capital to cover new vehicles, traffic would increase, "considerable savings will be made by co-ordination [of services]" and that they would, within a twelvemonth, show a satisfactory profit over and above the interest payable on loans.

How far this was realistic and how much 'pie in the sky' is doubtful for although the larger companies, especially PMT, were profitable, many of the smaller operators functioned only with the most basic returns; economies of the type where no wages were paid to the wife (albeit she was book-keeper, telephonist, booking office clerk and relief driver) being as common then as they were in the 1950s and 1960s. As Stoke and Newcastle could have outvoted the Rural Councils and were unlikely to be sympathetic to any loss making country route no doubt many of the 'Market Day Only' and inter-village routes would have been 'Beechinged'.

It has to be said, though, that O.C. Power, the traffic manager of the Birmingham & Midland Motor Omnibus Company - themselves a quasi-monopoly, thought differently, saying "We shall have the transport of a huge area controlled by 39 members of a board, each with a parochial interest to secure advantages for the area he represents and all fighting for their own ends". But then the BMMO (Midland Red) was 50% owned by railway companies (30% London, Midland & Scottish Railway, 20% Great Western Railway) and was a member of the same grouping as PMT (Tilling & British Automobile Traction), so Mr. Power may have had a (diplomatic) axe to grind!

When the Bill authorising this work was being promoted it was stated (1937) that the proposed Board had made provisional agreements to purchase at least 200 vehicles with a total expenditure of over £1,000,000 albeit this would include premises and equipment, and that they anticipated, in addition, buying 100 or more new vehicles. The date of acquisition was expected to be 1 August 1937, but the Potteries Motor Traction Company who owned around 250 out of the 500 estimated vehicles which would have been involved united with the railways and a number of local councils to oppose the Bill and it was not passed by Parliament.

As one of the Midlands transport might-have-beens it is fascinating (if futile!) to think how bus services might have evolved in the area, but whether co-ordination would have been completed prior to World War 2 is doubtful; during the conflict small and large operators alike served the area of the Six Towns well. One wonders if it is a relic of this plan that causes some inhabitants of Macclesfield to casually describe today's PMT vehicles as "Potteries Muck Trucks" for they were the loser having, as a matter of civic pride, planned to establish a municipal fleet during 1936/7 and being forced to abandon their approach to Parliament when the North Staffs Board's plans were announced.

HULLEYS OF BASLOW

The firm of Hulleys of Baslow was the ultimate country and Market bus operator, running many routes which, almost all the year round, lost money hand over fist but provided a service to the local population. To attain this ideal and still survive financially they made great use of a heterogeneous mixture of second hand vehicles. Towards the end, in 1978, this policy was no longer acceptable and they merged with another Derbyshire independent J.H. Woolliscroft & Sons Ltd (Silver Service). Hulleys were to be the fourth owner of No. 10 (JTB 965) since she was built in 1948. The chassis was unusual, a Crossley SD42/3 and the body even rarer, coming from TransUnited. Purchased December 1956, she was 'fettled' at Baslow, starting work in the Spring of 1957 and remaining in stock until September 1960. Even here, taken after repainting, the drivers cab door no longer fits and the emergency door had to be lifted to close it. The seats were somewhat unhappy and a 'bouncy' floor made the ride interesting. The draughts in the cab were phenomenal!

Horsley Woodhouse, three miles SE of Belper with a population (in 1931) of 1435, was home to three operators. Samuel Aldred was to become part of the Heanor & District trading group in 1934, although in May 1938 this consortium sold out to Midland General.

DERBY AND DISTRICT BUS DAYS

Within a few miles of Derby there was until the motorcar age a vast demand for both local and excursion transport. Certainly until the first world war (and in some remote areas long after) to get to market meant either a 5 a.m. or so departure on a cart laden with produce, a slow uncertain and agonizingly uncomfortable journey in winter however sylvan in summer, or one walked. Few of the population could afford a pony and trap and those that could were unlikely to visit the market much before the sun had risen.

Few motor-vehicles ran in passenger service with any real degree of reliability before 1910 and between then and 1914 even fewer people in Derbyshire could afford the (relatively) few pence fares charged.

After 1919 as stability returned so did the demand for travel but until 1921/2 this demand was still spasmodic - a trip to the market was fine, and a trip to 'Town' was quite reasonable on Saturday. And a nice char-a-banc outing - rarely more than 50 miles - was pleasant enough on a Sunday although today's nomadicism would have seemed frightening to most families in the 1950s let alone the 1920s and customers often had to be wooed with a 'Mystery' or 'Surprise' trip. Not too unsurprisingly, many coach or bus operators were also publicans; George Collis, Malt Shovel Inn, Aston-on-Trent; Harold Watson, Old Oak, Horsley Woodhouse; Bill (W) Liewsley, Vernon Arms, Spondon; Charles Yeomans, French Horn, Rodsley all offered some form of service. The link between horse-drawn carriers cart and motor-lorry was completed by entrepreneurs of this type, typically Charles Yeomans started with a Thorneycroft 4-ton lorry which he fitted with benches on a Friday and carried a mixed bag of passengers and animals to Derby market.

During the period prior to 1935 there were at least 55 bus and coach operators in the vicinity of Derby. Many of course were interchangeable, a bus becoming a coach in the evenings or on Sunday, a difficulty which may have led to the belief that a 'chara' ride is necessarily rather 'rough', a belief that lingers on today where some vehicle owners are so short-sighted that they use a 'boach' as a bus to transport schoolkids during the week, carry a booze outing on Saturday and then advertise the same vehicle (probably bearing the smell if not the evidence of the pub outing) for a "luxury tour" on the Sunday.

Of the 55 or so Derby and district operators, no less than 22 were to sell out to Trent Motor Traction, a round dozen to other 'major' companies, including Notts & Derby, Midland General and Bartons, and at least half of the remainder quietly died in the 1920s. The photographs, all from the collection of John Heath, do at least depict some of the vehicles used and the characters (often owner/drivers) who contended with poor roads, often irascible passengers, temperamental vehicles and poor financial returns.

E & H Frakes operated their green painted Chevrolet between Castle Donington & Derby in the 1920s. Pride of ownership is obvious.

The Gem service operated intermittently between Derby (Bold Lane) and Quarndon Park Nook from 1924 onwards. Its last runs were in 1932 when two Reo Speedwagons were in use, the newer (registered 9 April 1931) is seen here.

The Eagle service shown in this photograph was a joint one with that of H.S. North and ran from Derby to Heanor via Kilburn Toll Bar, this 1934 Morris Dictator being the normal vehicle used. Although S.O. Stevenson, the owner of Eagle operated many other services, he sold out to Trent in 1935. A curiosity was the reversible nameboard (Eagle one side, North the other) carried on RB 4393.

A 1927 Dennis of H.S. (Harry) North used on the Derby (Albert Street)-Heanor service. Another of the operators based at Hosley Woodhouse, Harry North painted all his vehicles in two shades of blue and, through judicious purchases of other companies, ended up with quite a complicated network. The Derby-Heanor service was sold out to Trent in May 1935.

CH 9136, first registered 12 April 1930, is the other Reo Speedwagon operated by F. Monks of 10 Curzon Street, Derby under the name of 'The Gem'. The differences between this model and that of a year later are obvious from the photograph of CH 9864 on page 42.

Although a poor quality photograph, nonetheless it is included as showing RA 5538 of J.T. Boam's Ray Service in 1928. The vehicle, a Leyland Lioness, is an extremely rare one to find in use on a country bus service. The Ray ran between Heanor and Ilkeston and other routes until bought out by Midland General in March 1931. Although not apparent here, livery was cream and green with gold lining out and black mudguards.

C.E. Salt, whose first operating address was 54 Derby Lane, Derby, ran a service from Derby to Belper via Duffield in 1919, utilizing a red Wolverhampton-built 'Star' bus, said to be that above. After moving to Tamworth House, Duffield, he expanded to run two 26-seater Gilfords and a Dennis 'Lancet', all this fleet being shown below. The leading vehicle (a Gilford) is fitted with Gruss air springs which acted as dampers on the front end and gave a very smooth ride. The timetable dated c.1930 is curiously clumsy in its layout, but its running time compares well with the 45 minutes required by Trent nine years earlier.

RED STAR SERVICE
DERBY & BELPER

	a.m.		F.L.		FSL	p.m.	FSE		FSL		FSL			F & S		S.O.	
Derby, Bold Lane		7 45	8 45	9 20	10 20	11 0	12 10	1 5	1 50	2 35	3 10	4 5	4 50	5 35	6 15	7 5	7 55	8 25	9 15	10 5
Darley		7 50	8 50	9 25	10 25	11 5	12 15	1 10	1 55	2 40	3 15	4 10	4 55	5 40	6 20	7 10	8 0	8 30	9 20	10 10
Allestree		7 55	8 55	9 30	10 30	11 10	12 20	1 15	2 0	2 45	3 20	4 15	5 0	5 45	6 25	7 15	8 5	8 35	9 25	10 15
Duffield	7 0	8 2	9 2	9 37	10 37	11 17	12 27	1 22	2 7	2 52	3 27	4 22	5 7	5 52	6 32	7 22	8 12	8 42	9 32	10 22
Milford	7 5	8 8	9 8	9 43	10 43	11 23	12 33	1 28	2 13	2 58	3 33	4 28	5 13	5 58	6 38	7 28	8 18	8 48	9 38	10 28
Belper	7 15	8 15	9 15	9 50	10 50	11 30	12 40	1 35	2 20	3 5	3 40	4 35	5 20	6 5	6 45	7 35	8 25	8 55	9 45	10 35

	a.m.					F.L.	S.L.	FSL		S.L.	FSL	FSL		S.L.			F&S	F.O.	S.O.
Belper	7 18	8 15		9 30	10 15	10 55	11 35	1 0	1 45	2 30	3 15	4 5	4 45	5 30	6 10	7 0	7 40	8 30	9 0	9 09	9 50	10 40
Milford	7 24	8 22		9 37	10 22	11 2	11 42	1 7	1 52	2 37	3 22	4 12	4 52	5 37	6 17	7 7	7 47	8 37		9 7	9 57
Duffield	7 29	8 28	8 50	9 43	10 28	11 8	11 48	1 13	1 58	2 43	3 28	4 18	4 58	5 43	6 23	7 13	7 53	8 43	9 10	9 13	10 3	10 55
Allestree	7 35	8 35	8 57	9 50	10 35	11 15	11 55	1 20	2 5	2 50	3 35	4 25	5 5	5 50	6 30	7 20	8 0	8 50		9 20	10 10
Darley	7 40	8 40	9 2	9 55	10 40	11 20	12 0	1 25	2 10	2 55	3 40	4 30	5 10	5 55	6 35	7 25	8 5	8 55		9 25	10 15
Derby	7 45	8 45	9 5	10 0	10 45	11 25	12 5	1 30	2 15	3 0	3 45	4 35	5 15	6 0	6 40	7 30	8 10	9 0		9 30	10 20

F.L.—5 mins. later Friday.   S.L.—5 mins. later Saturday.
F.S.L.—5 mins. later Friday and Saturday.
F.S.E—5 mins. earlier Friday and Saturday.
F. & S.—Friday and Saturday only.   S.O.—Saturday only.

Proprietor—
C. E. Salt, Duffield, Derbyshire.

Telephone—
Duffield 213.

WEEKLY TRAFFIC RETURNS AS PUBLISHED FEB, 11 1949

SYSTEM	Vehicle Type	1949 Week Ended	Receipts for week £	+ or − compared with corresponding week 1948 £	Number of Weeks	AGGREGATE TO DATE TOTAL £	+ or − to date £
BIRMINGHAM	Tramways Motorbuses Trolleybuses	January 9 - do - - do -	18,056 72,661 4,230	− 1,627 + 4,290 + 81	41 41 41	795,875 2,920,886 173,505	− 18,521 + 259,296 + 7,969
BURTON UPON TRENT	Motorbuses	January 30	2,287	+ 102	44	96,932	+ 6,922
CHESTERFIELD	Motorbuses	January 30	6,580	+ 60	44	283,863	+ 13,078
COVENTRY	Motorbuses	February 5	17,019	− 118	44	778,349	+ 86,231
DERBY	Motorbuses Trolleybuses	January 29	2,541 5,869	+ 416 + 828	43 43	112,246 265,875	+ 27,499 + 52,571
LEICESTER	Tramways Motorbuses	January 29	4,451 10,109	− 1,577 + 1,399	43 43	247,025 385,303	− 7,289 + 41,980
NORTHAMPTON	Motorbuses	January 29	4,872	− 90	44	216,797	+ 13,653
NOTTINGHAM	Motorbuses Trolleybuses	January 22	13,635 8,866	+ 655 + 235	42 42	570,848 382,150	+ 37,197 + 4,556
WALSALL	Motorbuses Trolleybuses	January 23	9,530 1.913	+ 525 + 79	43 43	405,438 79,896	+ 21,731 + 1,811
WOLVERHAMPTON	Motorbuses Trolleybuses	January 30	5,130 10,506	+ 578 − 218	44 44	229,162 479,190	+ 26,761 + 25,673

32 DERBY: Allestree (9); Duffield (18); **BELPER** (28).
LEAVE **Derby** (Bus Station Bay 13) **Mon-Fri** A.M. 7.28, 8.23, then hourly to 11.23; P.M. 12.23 then hourly to 4.23, 5.33, 6.03, 6.33, 7.33, 8.33, 10.10, 11.00. **Sat.** only: A.M. 8.23 then as **Mon-Fri**.

BELPER; Duffield (10); Allestree (19); **DERBY** (28).
Leave **Belper** (Chapel Street) **Mon-Fri.** A.M. 6.43, 7.40, 8.40, 9.45, 10.40, 11.40; P.M. 12.40 then hourly to 6.40 then 12.15 midnight. **Sat.** only: A.M. 7.40, 8.40, 9.45, 10.40, 11.40 P.M. 12.40 then hourly to 6.40 then 12.15 midnight.

Note: In addition Trent routes R1 30 and 31 cover the whole or part of this service; which is quoted, with permission from 'Trent Buses' to compare with the older timetables.

INTENSIVE WORKING

Notwithstanding a complex set of railway workings on the Nottingham-Derby route, Barton Transport merely responded by increasing the number of vehicles on the road. This 1948 timetable gives some idea of the demand faced and met in those car- (or least petrol-) less days. Return workings, omitted for clarity, were equally intensive.

ROUTE No. 5

DAILY SERVICE BETWEEN
NOTTINGHAM - DERBY
(MOUNT STREET BUS STATION)   (CENTRAL BUS STATION)

SUNDAY IMPROVED SERVICE

	A.M.														
											D	W	D	W	D
Nottingham Mount Street ..dep.	—	—	—	8 33	9 3	—	9 33	—	10 3	1018	1033	104	11 3	1118	1133
Beeston Square	6 6	7 0	8 14	8 44	9 14	—	9 44	—	1014	1029	1044	1059	1114	1129	1144
Chilwell Garage	6 8	7 2	8 16	8 46	9 16	9 46	9 46	1016	1016	1031	1046	11 1	1116	1131	1146
Depot Corner	6 10	7 4	8 18	8 48	9 18	9 48	9 48	1018	1018	1033	1048	11 3	1118	1133	1148
Long Eaton	6 18	7 12	8 26	8 56	9 26	9 56	9 56	1026	1026	1041	1056	1111	1126	1141	1156
Breaston Church	6 25	7 19	8 33	9 3	9 33	—	10 3	—	1033	—	11 3	—	1133	—	12 3
Draycott Square	6 29	7 23	8 37	9 7	9 37	—	10 7	—	1037	—	11 7	—	1137	—	12 7
Borrowash	6 35	7 28	8 43	9 13	9 43	—	1013	—	1043	—	1113	—	1143	—	1213
Spondon Lane End ..	6 40	7 32	8 48	9 18	9 48	—	1018	—	1048	—	1118	—	1148	—	1218
Chaddesden Lane End	6 45	7 37	8 53	9 23	9 53	—	1023	—	1053	—	1123	—	1153	—	1223
Derby Bus Station arr.	6 52	7 44	9 0	9 30	10 0	—	1030	—	11 0	—	1130	—	12 0	—	1230

	A.M.														
	W	D	W	D	W	D	W		D	W		D	W	D	W
Nottingham Mount Street .. dep.	1148	12 3	1218	1233	1248	1 3	1 18	—	1 33	1 48	—	2 3	2 18	2 33	2 48
Beeston Square	1159	1214	1229	1244	1259	1 14	1 29	—	1 44	1 59	—	2 14	2 29	2 44	2 59
Chilwell Garage	12 1	1216	1231	1246	1 1	1 16	1 31	1 44	1 46	2 1	2 14	2 16	2 31	2 46	3 1
Depot Corner	12 3	1218	1233	1248	1 3	1 18	1 33	1 46	1 48	2 3	2 16	2 18	2 33	2 48	3 3
Long Eaton	1211	1226	1241	1256	1 11	1 26	1 41	1 54	1 56	2 11	2 24	2 26	2 41	2 56	3 11
Breaston Church	—	1233	—	1 3	—	1 33	1 48	—	2 3	2 18	—	2 33	2 48	3 3	3 18
Draycott Square	—	1237	—	1 7	—	1 37	1 52	—	2 7	2 22	—	2 37	2 52	3 7	3 22
Borrowash	—	1243	—	1 13	—	1 43	1 58	—	2 13	2 28	—	2 43	2 58	3 13	3 28
Spondon Lane End	—	1248	—	1 18	—	1 48	2 3	—	2 18	2 33	—	2 48	3 3	3 18	3 33
Chaddesden Lane End	—	1253	—	1 23	—	1 53	2 8	—	2 23	2 38	—	2 53	3 8	3 23	3 38
Derby Bus Station arr.	—	1 0	—	1 30	—	2 0	2 15	—	2 30	2 45	—	3 0	3 15	3 30	3 45

		P.M.									M'NT	A.M.	
		D	W	D	W	D	W	D	W	D	W	D	W
Nottingham Mount Street .. dep.	Repeat	10 3	1018	1033	1048	11 3	1118	1133	1148	12 3	1218	1 12
Beeston Square	every 15 minutes	1014	1029	1044	1059	1114	1129	1144	1159	1214	1229	1 23
Chilwell Garage	Buses leaving at	1016	1031	1046	11 1	1116	1131	1146	12 1	1216	1231	1 25
Depot Corner	18 mins. and 48 mins.	1018	1033	1048	11 3	1118	1133	—	—	—	—	—
Long Eaton	past each hour	1026	1041	1056	1111	1126	1141	—	—	—	—	—
Breaston Church	via Wollaton Park,	1033	1048	11 3	—	1133	—	—	—	—	—	—
Draycott Square	Buses leaving at	1037	1052	11 7	—	1137	—	—	—	—	—	—
Borrowash	3 mins. and 33 mins.	1043	1058	1113	—	1143	—	—	—	—	—	—
Spondon Lane End	past each hour	1048	11 3	1118	—	1148	—	—	—	—	—	—
Chaddesden Lane End	via Dunkirk	1053	11 8	1123	—	1153	—	—	—	—	—	—
Derby Bus Station arr.	until	11 0	1115	1130	—	12 0	—	—	—	—	—	—

W—Denotes via Wollaton Park   D—Denotes via Dunkirk

MONDAY TO FRIDAY

	A.M.						W		D		W	D	W	D
Nottingham Mount Street ....dep.	—	—	—	—	—	5 48	—	6 3	—	6 18	6 33	6 48	7 3
Beeston Square	4 44	5 16	5 41	—	—	5 59	6 8	6 14	6 23	6 29	6 44	6 59	7 14
Chilwell Garage	4 46	5 18	5 43	5 45	5 55	6 1	6 10	6 16	6 25	6 31	6 46	7 1	7 16
Depot Corner	4 48	5 20	5 45	5 47	5 57	6 3	6 12	6 18	6 27	6 33	6 48	7 3	7 18
Long Eaton	4 56	5 28	5 53	5 55	6 5	6 11	6 20	6 26	6 35	6 41	6 56	7 11	7 26
Breaston Church	5 3	5 35	6 0	—	—	6 18	—	6 33	—	6 48	7 3	7 18	7 33
Draycott Square	5 7	5 39	6 4	—	—	6 22	—	6 37	—	6 52	7 7	7 22	7 37
Borrowash	5 13	5 45	6 10	—	—	6 28	—	6 43	—	6 58	7 13	7 28	7 43
Spondon Lane End	5 18	5 50	6 15	—	—	6 33	—	6 48	—	7 3	7 18	7 33	7 48
Chaddesden Lane End	5 23	5 55	6 20	—	—	6 38	—	6 53	—	7 8	7 23	7 38	7 53
Derby Bus Station arr.	5 30	6 2	6 27	—	—	6 45	—	7 0	—	7 15	7 30	7 45	8 0

	P.M.											
		D	W	D	W	D	W	D	W	D	W	W
Nottingham Mount St., dep.	Repeat	9 3	9 18	9 33	9 48	10 3	1018	1033	1048	11 3	1118	1148	12 3
Beeston Square	every 15 minutes	9 14	9 29	9 44	9 59	1014	1029	1044	1059	1114	1129	1159	1214
Chilwell Garage	Buses leaving at	9 16	9 31	9 46	10 1	1016	1031	1046	11 1	1116	1131	12 1	1216
Depot Corner	18 mins. and 48 mins.	9 18	9 33	9 48	10 3	1018	1033	1048	11 3	1118	1133	—	—
Long Eaton	past each hour	9 26	9 41	9 56	1011	1026	1041	1056	1111	1126	1141	—	—
Breaston Church	via Wollaton Park,	9 33	9 48	10 3	1018	1033	1048	11 3	—	1133	—	—	—
Draycott Square	Buses leaving at	9 37	9 52	10 7	1022	1037	1052	11 7	—	1137	—	—	—
Borrowash	3 mins. and 33 mins.	9 43	9 58	1013	1028	1043	1058	—	—	1143	—	—	—
Spondon Lane End	past each hour	9 48	10 3	1018	1033	1048	11 3	—	—	1148	—	—	—
Chaddesden Lane End	via Dunkirk	9 53	10 8	1023	1038	1053	11 8	—	—	1153	—	—	—
Derby Bus Station arr.	until	10 0	1015	1030	1045	11 0	1115	—	—	12 0	—	—	—

W—Denotes via Wollaton Park   D—Denotes via Dunkirk

SATURDAY

	A.M.					W		D		W	D	W	D	
Nottm. Mount St... dep.	—	—	—	—	—	5 48	—	6 3	—	6 18	6 33	6 48	7 3	Repeat every 15 minutes
Beeston Square	4 44	5 16	5 41	—	—	5 59	6 8	6 14	6 23	6 29	6 44	6 59	7 14	Buses leaving at
Chilwell Garage	4 46	5 18	5 43	5 45	5 55	6 1	6 10	6 16	6 25	6 31	6 46	7 1	7 16	18 mins. and 48 mins.
Depot Corner	4 48	5 20	5 45	5 47	5 57	6 3	6 12	6 18	6 27	6 33	6 48	7 3	7 18	past each hour
Long Eaton	4 56	5 28	5 53	5 55	6 5	6 11	6 20	6 26	6 35	6 41	6 56	7 11	7 26	via Wollaton Park,
Breaston Church	5 3	5 35	6 0	—	—	6 18	—	6 33	—	6 48	7 3	7 18	7 33	Buses leaving at
Draycott Square	5 7	5 39	6 4	—	—	6 22	—	6 37	—	6 52	7 7	7 22	7 37	3 mins. and 33 mins.
Borrowash	5 13	5 45	6 10	—	—	6 28	—	6 43	—	6 58	7 13	7 28	7 43	past each hour
Spondon Lane End	5 18	5 50	6 15	—	—	6 33	—	6 48	—	7 3	7 18	7 33	7 48	via Dunkirk
Chaddesden Lane End	5 23	5 55	6 20	—	—	6 38	—	6 53	—	7 8	7 23	7 38	7 53	until
Derby Bus Station, arr.	5 30	6 2	6 27	—	—	6 45	—	7 0	—	7 15	7 30	7 45	8 0	

	P.M. D	W	D	W	W	D	W	W	A.M. W	W	D	W
Nottingham Mount Street dep.	10 3	1018	1033	1048	11 3	1118	1133	1148	1155	12 3	1210	1225	1 12
Beeston Square	1014	1029	1044	1059	1114	1129	1144	1159	12 6	1214	1224	1236	1 24
Chilwell Garage	1016	1031	1046	11 1	1116	1131	1146	12 1	12 8	1216	1226	1238	1 26
Depot Corner	1018	1033	1048	11 3	1118	1133	—	—	—	1218	—	1240	—
Long Eaton	1026	1041	1056	1111	1126	1141	—	—	—	1226	—	1248	—
Breaston Church	1033	1048	11 3	1118	1133	—	—	—	—	—	—	—	—
Draycott Square	1037	1052	11 7	1122	1137	—	—	—	—	—	—	—	—
Borrowash	1043	1058	1113	—	1143	—	—	—	—	—	—	—	—
Spondon Lane End	1048	11 3	1118	—	1148	—	—	—	—	—	—	—	—
Chaddesden Lane End	1053	11 8	1123	—	1153	—	—	—	—	—	—	—	—
Derby Bus Station arr.	11 0	1115	1130	—	12 0	—	—	—	—	—	—	—	—

W—Denotes via Wollaton Park   D—Denotes via Dunkirk

3000 21743HL

OLD 602 (1044 in the Barton fleet) once previously carried London Transport colours as RTL 1493. Few purchasers of ex LT vehicles used the blind apertures to their full value.

This vehicle is one of the famous Barton rebuilds. Rebodied in 1981 it incorporated components from a Leyland TD1 and a TD5C. Classified BTS 1 bodywork was home-made. KAL 256, alongside, is a 1947 Duple bodied Leyland PD1.

BRUM'S BYGONE BUSES

No. 81, OL 1714 was built in 1923 with a John Buckingham 21 seat bus body on a Daimler CK2 chassis. Re-numbered 37 in 1931 for book-keeping purposes, it was withdrawn the same year.

3457, BON 457C was one of 2 Daimler Fleetline CRG 6 (Gardner-engined) single deckers bodied by Marshall of Cambridge in 1965, before the true single-deck Fleetline chassis became available. The objective in using this chassis was to have commonalty of spare parts with the double deckers in the fleet but this batch were the only ones to have the rear bulkhead and 'bustle' of their big sisters. Withdrawn January 1980.

90, OV 4090. One of a batch of ten Morris Dictator chassis built at the old Wolseley plant at Adderley Park, and fitted with Saltley built 34 seat all-metal bus bodies by Metro-Cammell in 1931. Only another four were bought by Birmingham City Transport (in 1933) and it is surprising that this one has survived, having been withdrawn in 1945. Photographed in the Midland Bus Museum, Wythall, May 91, as can be seen she awaits a sponsor so that complete restoration can take place.

254, OX 1520, is a superb example of a 1927 52 seater open-staircase body by Short Brothers of Rochester on an ADC 507 chassis. To modern eyes it is (probably) regarded as ugly and certainly there is a slight lack of streamlining, but they were considered up-to-date and reliable enough to operate the semi fast cross city service between the Maypole (Alcester Road) and Erdington (Chester Road) via New Street and Corporation Street. Withdrawn 1935.

493, OV 4493 was built just four years later than 254 and shows the evolution of the Birmingham double-deck omnibus. The chassis is an AEC Regent and the bodywork by Metro-Cammell of Saltley, seating rather surprisingly only 48 passengers. Around this period the B.C.T. appear to have been trying comparative bodies, similar 48 seaters coming from Brush, Short, English Electric and Vulcan as well as Metro-Cammell during 1929 and 1931. 493 was loaned to London Transport in October-November 1940 following the Blitz and withdrawn in 1945.

526, OC 526 carried a 1933 Metro-Cammell 48 seat body on an Adderley Park built Morris Imperial chassis. Seen here after withdrawal in 1942 this was not a happy class of vehicles primarily due to crankshaft problems and impossibly heavy steering. When a driver was rostered one of these he would find any reason to request a 'change' and it is said that windscreen wipers always 'fell off' when rain began and that strange noises were commonly heard by drivers, if not by fitters. Fifty were ordered by BCT out of 83 chassis made, withdrawal beginning as early as 1939 due to a shortage of spares, 526 leaving the fleet in 1945.

1401, FOP 401 was a member of the shortest lived class of vehicles purchased by BCT. The chassis was a Guy Arab II powered by a 5-cylinder Gardner engine and with a decidedly basic 'Utility' war-standard body by Strachans, whose products were always more angular than most. By the delivery date, 1945, other bodybuilders were beginning to soften the outlines but presumably Strachans panel beaters were still bashing aircraft panels and not curving bus bodywork. Withdrawn 1950.

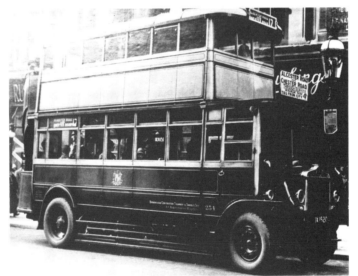

BY EASY STAGES
A PERSONAL RECOLLECTION OF A BLACK COUNTRY ROUTE
John Reohorn

Early summer. Roadside trees are bright with new foliage, the heathland, as yet undeveloped, is alive with herbage stirring in a gentle breeze. It is 1940 and war is far away; the prelude to Dunkirk still in its opening phases. We wait for a bus.

It comes into sight round the bend rolling purposefully past the houses. A single decker, still resplendant in its pre-war lined-out red, but even to infant eyes it seems extraordinary antique. It stops, the front entrance right behind the throbbing engine which is causing the whole vehicle to rock in sympathy. It stands high over its wheels and the steps seem more like a ladder; the lower ones have treads but no risers. From my eye level the chassis is visible. Short legs clamber into the angular body; the handrails are brass. A conductress presses against the bulkhead allowing us to squeeze past. At her signal the driver in his narrow cab engages the clutch: with healthy roar and whine of gears, the bus moves on.

Not an elegant carriage. Internally and externally there is a severity of line. The roof has a low arc; there is no indicator box. Illuminated numbers inside the window announce the route, 274. A painted board in slots declares the destination: Sedgley via Bilston. Oh how the company loved their Latin preposition. With equally typical ostentation the words 'Midland Red' are emblazoned in enamel across the width of the curiously Byzantine radiator.

The Black Country is a dense collection of distinctive towns packed cheek by jowl within an oval depression drained by the River Tame. It is shamelessly industrial. Sedgley, poised astride the southern rim where rural meets urban, tends to be aloof assuming the airs of a country town. Wolverhampton's green and yellow trolleybuses hissing towards Dudley challenged its ambition, but the Midland's single deckers somehow conspired with its pretentions.

This route began in Ettymore Road beside the Bull Ring, but once mobile they quickly plunged down off the ridge; a mile-long steady decline fringed by houses and grassy spaces. A good test of brakes.

Once across the 'New Road' (A4123), it is industry, that distinctive dereliction that follows abandoned collieries. Two railways cross and away to the left the sulphurous sprawl of Stewart and Lloyd's steelworks. Winding roads; erstwhile lanes masquerading as urban thoroughfares.

Traffic lights; right turn under wire again sharing the road with W.C.T. 25 between the strung out shops of Bilston. Past a palladian church and duck in beside the G.P.O. hard by the G.W.R. main line. Here the bus waits. It is hard-core industrial townscape yet only two and a half miles from Sedgley.

It was a feature of these urban routes to wait for significant periods at key points. Few passengers would take the whole journey; in reality many services were short routes strung together for operational convenience. It made for sound management and it suited the passengers well for there were regular connections and ample alternatives.

Back on the main road (A41) the bus is under wire again: WCT 2/7/47 as far as Moxley, not a lovely place. Here the trolleys bore away onto A4038 leaving the red saloon in sole possession of the main road, trundling past more factories and steelworks into the ancient borough of Wednesbury. Here the loading point was beside a derelict factory in High Bullen. Another wait, humbly, behind the FEDD's from Dudley while Walsall's blue 37s rattle past.

Here, five miles out, the bus could turn back using the roundabout at the top of the road. In that case its route number would be 273 and it would load beside the Hippodrome, where the incline dictated wheel chocks for safety.

Most went an extra mile, as 274. This took them through housing estates, the scene of my 1940 encounters, to Fallings Heath (There was a working farm here until c.1950). To turn, the bus was driven into a side road, reversed into a cemetery gateway and driven forward to load by the few shops. Conductors had whistles for signalling during the reversal.

A few buses went a further half mile where they contacted A4038 again, but with Walsall's 37/8 now in residence. To turn involved the bus darting out to make a wide sweep on the A-road around a bollard. Officially still in Fallings Heath, the company called this James Bridge and assigned 275 to the service. The loading stand was beside the main gate of F.H. Lloyd's steel foundry.

Less frequently still some buses would continue across the A4038 as route 276 penetrating to the heart of vibrant industry at Darlaston Green. Heavy engineering: great drab factories whose unrelieved brick walls stretched hundreds of yards at a time, where the very ground quaked from the beat of huge forges and presses, and the air was sharp with pungent scents of hot metal. Awe inspiring. Within this maelstrom of noise, route 277 joined up.

At evocatively named Bug Hole Bridge they broke free into open space, crossed the L.M.S. and slipped round Shepwell Green to enter Willenhall by its back door. They loaded by the fire station just beside another trolleybus route, the joint Walsall/Wolverhampton 29.

Although predominantly a single-deck operator B.M.M.O. rarely used them on its Black Country network. The 273-6 composite was an exception, but it was by no means a monotonous route. Indeed the rolling stock was quite diverse.

In the 1940's and 50's, when I used it, the mainstay was varieties of the 'ON' type, mostly DON variants. Nice useful buses: comfortable, reliable, well proportioned and well finished. The slightly newer SON was very stylish indeed, especially the small batch with Brush bodies. Within this basic diet could be found a healthy sprinkling of older exotic types. IM4's were quite common, and M's were also seen. I cannot be certain about my 1940 encounter, it could easily have been either an M or a QL. Wartime exigencies kept several of these veterans active into the fifties.

Old or new, they each had character and projected dependability. The company buses, with their moquette upholstery always seemed classier than the rexine clothed municipals. We forgave them their unreadable route numbers.

Post war, things began to change; new hands took the helm and we got roller blind destinations. There were also underfloor engines, forty seats, and heaters! The S6 was the standard type, a fine vehicle. It had probably the best standard of internal finish ever achieved by the company. It and the essentially similar wider S8's were the commonplace, they and their descendants. Eventually the day came when the appearance of a tatty SON was a nostalgia trip.

The route has changed dramatically. Roads obliterated, factories razed. Darlaston Green no longer quakes and few work there anymore. Nationalisation disembowelled the ailing B.M.M.O.Co. and deregulation has brought interlopers in bewildering quantity. Admittedly the pole which stands where F.H. Lloyd's gate once was bears the name Midland Red West, but the tired old National Mk I will not be red and it will not climb up to Sedgley. They call it rationalisation.

© John Reohorn 1992

Service No. 273 con.—SEDGLEY, BILSTON, and 1

			Service No.	274 PM	275 PM	SUNDAYS—continued.		274 PM	275 PM	
Sedgley (Ettymore Road)	..	..	dep.	5 47	6 17	....	....	6 47	7 17	....
Hurst Hill (Gate Inn)	..	approx.	,,	5 50	6 20	....	....	6 50	7 20	....
White Horse	..	..	,,	5 52	6 22	....	....	6 52	7 22	....
Bilston (Memorial)	..	..	,,	6 5	6 35	....	....	7 5	7 35	....
Wednesbury (Union Street)	..	..	,,	6 18	6 48	....	....	7 18	7 48	....
Myvod Estate (Beebee Road)	..	..	,,	6 23	6 43	....	....	7 23	7 53	....
James Bridge (F. H. Lloyds)	..	..	arr.	6 26	....	....	....	7 26	....	....

MIDLAND RED IN THE COUNTRY

Midland Red were always excellent at promoting local country services, partly to fill empty seats, but also to offer a quite inexpensive way of seeing some of the most beautiful English countryside. Luckily this tradition is continued by the successor companies, not the least by Midland Red West. For example, they give a 'General Guide' to both Market and Early Closing days. Markets are still quite widely held: Bishops Castle is on a Friday, Cleobury Mortimer on a Wednesday, Malvern a Friday, but Ludlow is worth a visit on Monday.

THE MIDLAND "RED" MOTOR SERVICES

'Buses run at frequent intervals from

THE BULL RING

NEW ST. & OLD SQUARE

BIRMINGHAM

to all the Towns, Villages, and Places of Interest in the Midlands.

200 ROUTES IN 16 COUNTIES

Time Tables can be obtained from all Conductors

Special Saloon 'Buses and Chars-a-banc for Private Parties of any description

WRITE FOR QUOTATION

O. C. POWER, *Traffic Manager*

CHIEF OFFICES

Bearwood	Birmingham
2577 Midland	Bull Ring, Central 386

Country bus of the 1930s & 1940s. Midland Red 2025, an SOS SON with English Electric built 38 seat bodywork, built in 1938. Incredibly 2025 survived in service until 1956!

Country bus of the 1950s & 1960s. 4701 was a Carlyle Works built S14. She entered service in 1958, and was withdrawn in 1970.

Country bus of the 1980s & 1990s. On an almost identical route to 4701, 669 (RDA 669R) a Leyland Leopard with Plaxton 49 seat coachwork built in 1981, awaits departure from Hereford on the run to Worcester March 1992.

READY FOR WORK

Although the name of 'Midland Red' still survives the new companies bear very little resemblance to the Birmingham & Midland Motor Omnibus Company. The old 'Midland Red' specialized in advanced designs which were the forerunners of today's vehicles, albeit neither bus drivers nor bus passengers enjoyed the most comfortable of vehicles, a 1950s doorless vehicle was not the best of machines in the middle of winter on the long runs out to Shrewsbury or Coventry! On the other hand their coaches were among the best available and the crews certainly did their best to uphold the traditions of British coach tours.

Despite an engine date of 1927 this chassis is one of the first of the "M" class (possibly No. 909/HA 4909) of 1929. Known as the 'Madam' class, softer springs were fitted than hitherto and a rather less spartan interior was called for from the bodybuilders. The use of the magic letters 'SOS' on every exposed surface is not, as is commonly thought, the reaction of a London driver seeing one of these vehicles but, probably, stood for "Shires Own Specification" or "Shires Omnibus Specification", L.G. Wyndham Shire being the engineer-in-charge.

The vehicles of the "Red" were once described by an embittered private operator as being "bloody ubiquitous". One saw what he meant, for not only did they appear in their rightful home, the Midlands, but wherever one went on a tour, or even a day trip, hundreds of miles from Edgbaston, there would be another Carlyle Works vehicle.

PAPER SHORTAGE.

IMPORTANT INSTRUCTIONS TO CONDUCTORS.

1. It will not be possible to reprint the Fare Book for some time; but occasional changes in fares will be notified by means of slips which must be carefully affixed to the appropriate pages, thus keeping the book up to date. On no account must pages of the fare book be detached, and

2. Fare Books must be handed in by Conductors on termination of their employment with the Company, for the use of their successors.

Roger Kidner, transport historian and publisher, was in Atherstone during August 1930 and caught this SOS FS built with Brush 34 seat bus body, at an angle that shows just how high and gaunt these vehicles were. The destination board reads "Coventry via Atherstone & Nuneaton".

A year later (3 September 1931) and the photographer was in Stratford-upon-Avon at a location that can still today just be recognised. The vehicle would seem to be of the 1927 "Q" (Queen) class.

412, FHA 412, was of class ONC with much more modern looking Duple bodies delivered in 1939. 30 seaters, their blinds were much more functional (albeit here disused), although the trim still takes away any severity in appearance. Seen here at Cheltenham in 1952 waiting departure for Birmingham.

2470 CH 9925 was quite an adventurous vehicle. Built 1931, to class IM4 with a Short Brothers 34 seat body, she was new to Trent Motor Traction. She was taken over by the War Department in early 1940 and then passed on to Midland Red in 1941/2 as part of a consignment of 44 vehicles. She re-entered service after refurbishment in 1942 and was withdrawn in 1947.

A Duple bodied class C2 coach shows the similarity of line to FHA 412, albeit this machine, one of a batch of 12, was not built until 1950. The two other vehicles on show on the card are a Burlingham bodied Leyland of Harrisons (Morecambe) and an ECW bodied Bristol tourer belonging to the Scottish Motor Traction Group. The sheep were there long, long, before motor vehicles.

2012, CHA 994, was bodied by English Electric in 1937, and positively oozes the charm of the 'Odeon' school of 1930s architecture. In many ways this is a quite beautiful coach from her 'Art Deco' grill to the broken window line. Class SLR (Saloon, Low, Rolls Royce).

UHA 253, fleet number 4253 must serve us as a memory of the Midland Red class of coaches that carried so many passengers on the famous 1950s "cruises" (evocative word!). Honeymooners, courting couples or lonely people, all graced the seats of the eleven 1954 built C4 vehicles. Seen here at Penzance Coach Station August 1961, 4253 carried her Alexander built 32 seater body until the end in 1966.

MIDSCENE - STRATFORD-UPON-AVON

On 2 August 1991 John Wilson, a BR locomotive driver-cum-enthusiast took himself and his Yashica T2 Compact camera to Wood Street, Stratford-upon-Avon, where he spent two hours in pleasant photography. The selection is drawn from a choice of 30 different vehicles or liveries. Others appear elsewhere in the book.

No. 952 Leyland (ex Daimler) Fleetline. MCW (Saltley-built) body, 73 seats, built 1976, so-called "Londoner" class for London Transport No. DMS 2044. Midland Red(S) livery.

No. 961 The last true Leyland design. Olympian with ECW 72 seat body built 1984. New to the company. Maybe in view of recent events Leyland flew too near the Gods in Olympus! Stratford Blue livery.

No. 950 Daimler Fleetline with Scottish built Alexander 77 seat bodywork delivered 1971 to the Trent Motor Traction Company, their number 945. Midland Red (S) livery.

No. 474 of a type known as 'breadvans'. Freight Rover Sherpa 365 with Rootes 16 seat body new 1987. Stratford Blue livery.

No. 302 Mercedes Benz 709D with PMT (Stoke-on-Trent) 25 seat bodywork new 1989. Midland Red (S) livery.

No. 584 Leyland National Mk.I, 49 seat built 1977 and ex 'Midland Red' Omnibus Company. Named 'Rosemary'. Stratford Blue livery - her sister 581 coming up behind is in Midland Red (S) livery.

TRANSPORT AT WAR JULY - DECEMBER 1942

During the Second World War transport operators saw all the aspects of human activity. Buses and trams were bombed, machine gunned and too often involved in accidents due to poor, restricted, lighting with inevitably some fatalities among the crew. Busmen and women alike often went on duty after a night in the shelter (where you did not get much sleep) and came back after 10, 11 or 12 hours not knowing if they would have a house to return to. Everything was rationed and spare time limited.

Passengers could be miserable, harassed, unpleasant or, with drink in them, violent. A handful of conductresses were raped walking home after they'd put away the last bus, while others could find themselves escorted home by a Serviceman which friendship occasionally led to marriage. It was not permitted for war news to be bad, although by 1942 there was some reduction in censorship. But also in 1942 fuel, rubber and vehicles were all in desperately short supply. German submarines, newly fitted with Metox Radar detection equipment (the so-called 'Bombay Cross') were to sink a greater tonnage of Allied shipping in that first six months than at any other time before or subsequently, some 2,500,000 tons including 142 tankers. The following notes are drawn from the material relevant to the Midlands that was published in "The Transport World", plus I have added a few salient notes on activities in other parts of the Kingdom that had a relevance to Midlands operations.

Judy-Joan Wright

BUS DRIVER ABSENTEE FINED
Summonses Under New Procedure

A 21-years-old Irish bus driver employed by the Trent Motor Traction Co., Ltd., who pleaded guilty to absenting himself from work on three occasions was fined £5 on each count—£15 in all—at Derby. The alternative was 75 days' imprisonment.

The driver said, after his conviction, that he came to this country from Ireland for six months, and when he wanted to go back he was not allowed to do so because his passport was not in order. He was anxious to go back there.

On hearing this statement, Alderman G. Wood, chairman of the bench, said the fine must be paid the same day or the order for imprisonment would operate.

Mr. C.F.R. Cleaver, for the Ministry of Labour, said the cases were brought under the Essential Works Order, and though there were three summonses only, this driver had been absent and late on other occasions. Under the new procedure, before the summonses were brought the matter went before the works committee consisting of representatives of the employers and the man's own workmates.

Mr. G.C. Campbell-Taylor, general manager of the Trent Motor Traction Co.Ltd., in evidence said that between February and July 6 the driver had been absent 32 times and late seven.

Alderman Wood said this was a very serious case in view of the shortage of labour.

WOMEN BUS DRIVERS
On Nottingham Routes

The first woman to drive a public service vehicle for a full shift for Nottingham Corporation transport department has recently completed her training as a trolleybus driver. At present she is employed as a conductress, but she will soon be going on to full time driving. Before the war she worked in a gown shop, but she could drive a car, and enrolled as an ambulance driver on the outbreak of war. She has been a conductress for over two years.

Six more women are being trained in driving by the department.

October Road Accidents

During October last 697 people were killed on the roads, compared with 857 for October, 1941. Persons seriously injured numbered 3,369, and persons slightly injured 9,849. During black-out hours 285 persons lost their lives, a big increase over September black-out deaths, which totalled 158. Deaths in the black-out during this month since the beginning of the war have, however, steadily decreased.

Conductor Tom Jones (Sgt.Pilot) is still at your service

But he doesn't call "Fares please" now. The passengers in his present bus do not pay fares. As a matter of fact one would hardly call them passengers when they're dropping four-thousand pounders on appropriate spots. So instead of "Fares please", it is "B for Bertie, B for Bertie calling" now, and happy landing, we hope, at the end of it all.

Tom Jones is one of many of our men who have left the platform or the driver's seat or the office stool to do another job. They've left us to carry on with depleted and substitute staffs. And they would be the first to say - knowing the difficulties and the additional special workers services - that we are not making a bad job of it either - thanks to the ladies.

Travelled Without Payment

When a man was summoned at Nottingham recently for failing to pay his bus fare, it was stated that he sometimes travelled on Corporation buses for a week at a time without paying a fare. He pretended to have a pass, but did not show it. He was fined two guineas.

Birmingham Transport's Losses by Theft

When two conductors were each fined £10 for stealing from Birmingham Corporation, a police sergeant stated that the City transport department was probably losing £500 a week as a result of thefts.

CHRISTMAS EVE TRAVEL AT COVENTRY
Appeal to Shoppers

An advertisement in the local press contains an announcement by Mr. R.A. Fearnley, general manager of Coventry Corporation transport department, appealing to shoppers to avoid Christmas Eve travel. It avoids the formal language of official announcements, and the style of appeal should find a ready response. It reads:-

"In peacetime our object would be to fill the streets with brightly-lit buses. This year we must be grim and gay with the ordinary services, and no extras. Please shop a week early, or the queue will be too long."

Bus Used for Wedding Party

A summons against Associated Bus Companies Ltd., of Hanley, for unlawful use of motor fuel, followed the use of one of the company's buses to convey a wedding party from church to a reception. The case was dismissed after Mr. Albert Beech, director and secretary, had pointed out that on this particular journey of fifteen miles one and a half gallons of diesel oil at most had been used. If taxis had been engaged for the journey they would have taken at least five times more fuel of a higher grade.

REDUNDANT MOTOR COACHES
Use Being Made of Spare Materials

In the House of Commons on September 30, Major Lyons asked the Minister of Supply whether he could estimate the number of motor coaches from all services lying redundant or derelict at depots throughout the country; what proportion had had good tyres removed, or had been jacked-up for tyre protection; how many had been returned to their former owners; and what steps were proposed to ensure the rapid liquefying of the vehicles before their deterioration was complete.

Sir Andrew Duncan replied that, in addition to 306 coaches returned to operators, and 157 others offered to their former owners, 658 coaches notified by Service departments for disposal had been transferred to other government departments or broken down for spares. Of the 616 remaining, many had only recently become available for disposal. In some cases it had not been possible to identify the former owners from the impressment records, and the Ministry of War Transport were considering whether, after expiration of a limited period, the vehicles should be offered to operators approved by them. Under that arrangement the vehicles should soon be disposed of.

Christmas Day Buses at Nottingham

Nottingham City Transport Committee has decided upon its Christmas Day bus services, subject to the approval of the Regional Transport Commissioner, and grant of fuel. The service — of petrol buses only — is to run from 10 a.m. to 4 p.m. at ordinary single fares. Any necessary works services will be run.

Control for Public Service Vehicles

The Ministry of War Transport has announced that all public service vehicle operators owning fewer than thirty vehicles must in future buy their tyres through authorised tyre depots, within a ten-mile radius of the place at which the vehicles are based. This new ruling, made in response to an urgent request from the Tyre Control, reverses a previous decision.

Coventry Loading Barriers

Coventry Corporation transport department has erected over 60 loading barriers for passengers, and it is intended to erect more as materials become available. They have been found a great help in the formation of queues.

NOTTINGHAM TRANSPORT
First Two Women Trolleybus Drivers

Nottingham's first two women trolleybus drivers have completed their 100 hours of dual service, and are ready to go on the road alone. They will be listed as spare drivers, ready to take up duty when required, and meanwhile acting as conductresses. Both were recruited as volunteers from the City Transport Department's 578 conductresses.

BUS ENGINE LEFT RUNNING
Walsall Fuel Wastage Case

At Walsall last week a Corporation bus driver was summoned for wasting fuel by leaving his engine running.

A special constable described how he boarded the bus to return home for lunch. After waiting for some time he realised that the engine was running. He alighted, and pointed this out to the driver, who told him that if the engine was stopped it would need three men to start it again.

Appearing for the driver on the instruction of the Transport and General Workers' Union, Mr. A.V. Haden pleaded that more fuel would have been wasted if the engine had been switched off. Mr. Haden said that, despite an instruction from the general manager not to leave the engine running, Walsall Corporation transport drivers had been advised by other responsible persons not to switch off because of the difficulty in restarting.

He added that 75 per cent. of Walsall buses had engines without starting handles, and there were special emergency depots with men continuously on duty to tow in buses which could not be started again, and that they had to do this two or three times a day.

The driver stated that he left his engine running deliberately to economise. He said he was "as patriotic as anyone and would not waste petrol".

He was cautioned and ordered to pay 19 shillings special costs.

Lateness at Birmingham

When two Birmingham conductresses were fined 60s. each for being persistently late, they made the excuse of having no alarm clock. It was stated that the persistent lateness of conductresses employed by the City transport department necessitated the calling every day of 25 per cent. more than were required, to ensure that the early morning buses were available for workers.

Coventry Old-Age Pensioners' Travel

Coventry City Council asked the Transport Committee to prepare a scheme allowing free travel by old age pensioners, the cost to be defrayed by the rates. The Transport Committee has reported that such a scheme would be impracticable. There were about 941 "non-contributory" old age pensioners in the city, and an unascertained number of "contributory" pensioners. If they were to have free travel it would have to be restricted to non-peak periods, and the annual cost to the Corporation would be £6 10s. per person. Furthermore, such a concession would lead to similar applications from other classes of the community.

Trafficking in Priority Permits

The Nottingham City transport department finds that there is trafficking in fire-watchers' priority bus permits. People queueing in the black-out for late buses have been approached with offers of priority travel permits on payment of 1s. or 1s.6d. Measures are being taken in the issue and cancellation of priority tickets to make it more difficult for people to carry on such practices.

EARLIER TIMES FOR LAST VEHICLES
Commissioner's Appeal to Birmingham Corporation

Sir Arnold Musto, Regional Transport Commissioner for the West Midlands, has asked the Birmingham Corporation to run the last buses in the city at 10 p.m.

"These late buses, running after ten o'clock in the evening," he says, "cater mainly for cinema-goers and pub-crawlers. That is something we must try and prevent".

At present the last city bus runs at 10.30 p.m.

Sir Arnold pointed out that in Coventry no buses run later than 9.0 at night, and spoke of the increasing seriousness of the tyre situation.

Leicester City transport department has arranged for last buses to leave the city centre at 10.15 pm. and 10.20 p.m. during the winter months.

CURTAILMENT OF MIDLAND BUS SERVICES
Public Urged to Co-operate

Mr. H. Mayo, chief assistant to Mr. J.H. Stirk, North-Midland Regional Transport Commissioner, is asking the public to co-operate in meeting the new conditions when the bus curfew - 9 p.m. for large towns and 8 p.m. for small towns - becomes operative on November 14.

He considers impracticable the suggestion that special services would be provided for fire watchers, Home Guards and civil defence workers coming off duty on Sunday mornings, when Sunday morning bus services are to be discontinued.

Mr. J.H. Stirk, in announcing the restriction, said he considered that Sunday services could be stopped entirely with the exception of workers' special services and services between towns, running for a short period, say, between 4 p.m. and 8 p.m. For the time being, however, he was only going to cut out the Sunday morning services.

In this connection it is to be noted that Coventry in the Midlands region is prepared to make arrangements for Sunday services conforming to the requirements adopted by the Regional Transport Commissioner for that area, and not, as previously stated, to discontinue Sunday services altogether, unless such a measure becomes general.

Supplementary Clothing Coupons for Transport Workers

The Board of Trade has arranged that 10 supplementary clothing coupons shall be available for workers of either sex regularly engaged on certain types of labour for an average of not less than 22 hours a week. This arrangement covers workers engaged in the operation of buses, coaches and road goods vehicles; garage workers; those engaged on construction, repair, scrapping and maintenance of buses, coaches and road vehicles. The extra coupons will be issued through the Ministry of Labour and National Service in the form of Supplementary Coupon Sheets (S.C.5a), containing 10 coupons each.

The shopping hours

It may mean breaking a good habit, but whilst the war is on, please don't use the buses for shopping when the workers need them. They *must* get to their work on time, and it isn't fair to delay their journey home after a heavy shift. After all there's plenty of time, when the workers are not travelling, for you to do your shopping. Don't make the workers walk!

Accidents in May

The figures issued by the Ministry of Transport for May show a decrease in the number of road fatalities compared with that of the same month last year; 505 persons were killed, compared with 701 in May, 1941. Black-out deaths numbered 60, compared with 150 last year. During other hours, 445 persons were killed. People seriously injured numbered 2,780, and people slightly numbered 8,398.

Coventry Protest Against Women Fire-watchers

A letter has been sent to the Home Secretary from 357 male and female employees of Coventry Corporation transport department. They state that they "view with grave apprehension the Order for compulsory fire-watching by women both in business premises and residential quarters in 'target areas.' " They "believe that this . . . Order is an outrage to society, that it is revolting to all aspects of masculine citizenship." They request Mr. Morrison to withhold the operation of the Order.

VICTORIA COACH STATION TO CLOSE

Following the order of the Minister of War Transport under which long-distance coach services are to be withdrawn from the end of September, Victoria motor coach station is to be closed. The station cost over £250,000 when it was opened in March, 1932, by London Coastal Coaches, Ltd., and in normal times it was used by the buses of about 50 operating companies.

Employees' Absenteeism

For being late on 51 occasions out of 336 days and absent without excuse on 43 days, a tram conductress employed by the undertaking was fined 40s. on each of three charges, two for being absent and one for being persistently late, at the City Police Court last week.

For the prosecution it was stated that on one day there were between 350 and 400 of the department's employees absent owing to sickness or other causes, and the services of 100 people from the department's offices had to be enlisted to perform the duties of the absentees in addition to their own.

STAFF PROBLEMS AT LEICESTER
Helped by No Sunday Morning Travel

Following protests against the suspension of Sunday morning tram and bus services in Leicester, the transport department state that, although the restriction was imposed in the interests of national economy, they view it as a solving of their staff difficulties. The change gives the department the added week-day service of 134 more people.

Sickness amongst the employees is causing some trouble. On some mornings as many as 110 journeys are lost through absences, and there has been a shortage of staff for over 12 months. About 750 girls have been trained since women conductresses were first employed, but the actual number now working is 350. Many conductresses have been unable to stand the strain, or have left through domestic reasons.

MIDLAND "RED" WAR COLLECTIONS

Up to the end of October, 85,839 National Savings Certificates had been purchased by employees of Midland "Red" Motor Services. These represent a cash value of £64,579 5s. During the month 4,851 tickets were purchased. The Regional Industrial Officer has congratulated them on their achievement.

The directors of the Midland "Red" have sent 100 guineas to the Red Cross and St. John War Organisation. The employees have contributed £8,000 to this fund.

COVENTRY'S PRESS ADVERTISING

Coventry Corporation transport department is making good use of press advertising in its appeals for the co-operation of the public.

Well-displayed advertisements have appeared in the *Coventry Evening Telegraph* this week. One urges former car users, now deprived of their basic petrol ration, to avoid travelling on buses at rush hours, whenever possible. Another is a request to drivers of road vehicles to avoid parking close to a bus stop, as this hinders the bus driver from stopping in his proper place, and the queue of passengers in boarding the bus.

These are excellent examples of how transport undertakings can use the local press in getting public support.

Long-distance Bus Cuts in Midlands

United Counties Omnibus Co. Ltd., have suspended their services from Kettering to Cheltenham and from Northampton to Oxford, on the instructions of the Regional Transport Commissioner. These did not duplicate train services, and their retention was supported by local authorities of the towns on the routes served.

The fact that goods made of raw materials in short supply owing to war conditions are advertised in this journal should not be taken as an indication that they are necessarily available for export.

Wolverhampton's Women Trolleybus Drivers

The first two women trolleybus drivers attached to Wolverhampton Corporation transport department have been passed out for driving. Both were formerly conductors. One of them, Miss Amy Davies, was previously a domestic servant, and last week broadcast in the one o'clock news her experiences in learning trolley bus driving.

Women Trolleybus Drivers

Newcastle Transport and Electricity Committee has decided against the employment of women trolleybus drivers. The undertaking uses double-deckers carrying about 70 passengers, and it is considered that the job of driving them would be too risky for women.

No Sunday Travel

Coventry Corporation transport department, besides accepting the regional pronouncement, intends to discontinue all transport facilities for the general public on Sundays from the end of November. Special facilities will be provided for war workers.

Bus maintenance staff — otherwise engaged

There's a deal of difference between a forward repair unit and the steady, daily routine of bus maintenance shops. But the fellows have taken to it like ducks to water — quite a lot of them. Today the buses are harder pressed than ever — new special services for workers, in all weathers and black-out — with depleted and substitute staff. It means, of course, inconvenience at home and buses not so spick and span as they used to be, but the country is fortunate in finding such men, ready trained and eager to do a job of work that ranks so high in the tactics of mechanical warfare.

Cuts in Birmingham Services

Cuts in Sunday transport services are anticipated in Birmingham soon, as it is said that at certain hours of the day the buses are overloaded with people travelling for pleasure. Curtailments have already been effected both by the Midland "Red" and Birmingham City transport department. The Midland "Red" last bus runs in the city at 9 o'clock, and in the country at 8 o'clock. Summer services are ruled out entirely, and long-distance buses are no longer entered upon the time-table.

TRANSFER OF CONDUCTRESSES
Portsmouth Resists Proposals

The Ministry of Labour and National Security and the Regional Transport Commissioner have come to an agreement to transfer ten conductresses from Portsmouth Corporation to similar employment in Oxford. They have informed the Portsmouth undertaking that a number of single women in their employ must be interviewed to determine whether they are "mobile".

Provided time is allowed to train substitutes, the Ministry considers that three conductresses a week might be drawn from Portsmouth to satisfy Oxford requirements. Suggestions are also made for transferring to Oxford conductresses from other undertakings.

Portsmouth Passenger Transport Committee has protested and instructed the general manager, Mr. Ben Hall, to resist the proposal. Councillor H. Lay, chairman of the Committee, states that the matter is to be referred to the Municipal Passenger Transport Association, and, if necessary, he is prepared to have a question asked in Parliament.

"We have trained these girls", he said. "Some have been with us over two years. It is most unfair that they should be taken away for similar duties elsewhere".

Some of the girls have been interviewed at the Labour Exchange, and when they have protested they have been threatened with transfer into the Services or munitions.

The Oxford services are operated by the City of Oxford Motor Services Ltd.

Leicester Churches and Bus Curfew

Leicester church leaders have estimated that the withdrawal of transport facilities on Sunday mornings will lose them half of their congregations and hundreds of pounds income. They say that church fellowships will be broken and that collections for charities will suffer. Leicester Christian Council has made a protest to the Ministry of War Transport. Mr. Noel-Baker said in Parliament on Wednesday that the position was being reviewed.

CONDUCTOR SHORTAGE IN COVENTRY
Paid and Auxiliary Workers Wanted

Coventry Corporation transport department is advertising for women to act as full-time or part-time bus conductors. Hours of work for part-time conductors are from 6 a.m. to 9 a.m., and 4 p.m. to 7.30 p.m.

The department is also inviting volunteers to act as auxiliary conductors during peak hours. They will be given a badge and a card of instructions, and asked to take charge of the platform of the bus while the "regular" conductor is collecting fares. They will be responsible for loading arrangements, and will give the signal for the bus to restart. While on duty they will travel free.

It is thought that they may also be able to assist in directing the formation of queues at busy stops.

Mr. R.A. Fearnley, general manager of the undertaking, has stated that about 500 auxiliary conductors will be needed. Their services will be most valuable between 6 and 9 a.m. and between 5 and 7.15 p.m.

CONVERT THOSE SLOTTED SEATS

In Five Hours!

The complete conversion can be carried out in five hours—either by your own fitters or at our works by appointment. The units are supplied complete, with full instructions for fitting.

ROWLAND HARTRICK LTD.
WARWICK ROAD · BIRMINGHAM · 11
'Phone: VICtoria 3048   'Grams: Victoria 3048, B'ham

Why Wooden Seats?
DIFFICULTIES OF ENSURING COMFORT
by A.W. HAIGH, A.M.I.Mech.E., A.M.I.A.E.

With the reintroduction of wooden seats into the passenger transport world (so that moquette, hides and other essential upholstering materials may be saved) the question of accommodating passengers comfortably with such an inadequate foundation is both urgent and difficult of solution.

A wooden seat, no matter for how short a time it may be occupied by the passenger, can be a source of acute discomfort unless it is designed to meet conditions arising from the constantly changing speed of the vehicle and the varying road surfaces over which it must travel. Most of us can remember the results of sitting on the slatted reversible seats on the older tramcars; the jolts caused by every rail joint being transferred to the passenger by the rigid timber, the partial paralysis set up by its continued unyielding pressure on the body and the always prevalent tendency to slide off the slippery surface under the influence of one's own weight, deceleration and even the very low centrifugal force of a cornering tram.

Shock-Absorbing Qualities Needed

The conditions obtaining in a tramcar, with perhaps one exception, are greatly exaggerated in the modern bus. It is true that at low speeds comparatively small road imperfections are accommodated by the tyres, and that only slight vertical acceleration is transferred to the passengers through the springs. As speed increases, however, even the minor obstructions met with on city streets and suburban roads cause sufficient spring deflection to set up an appreciably rapid rise and fall in the vehicle body, which in turn is passed to the passengers, so that, unless the seats possess adequate shock-absorbing qualities, their occupants are subject to all the jars set up by spring rebound.

Because of the higher operating speeds of buses, roll is more pronounced, but, because of the longer wheelbase, pitch is not so noticeable. But even the latter phenomenon of pitch, though indisputably reduced, does occur, as there is only one road speed which gives a relationship between the deflection and periodicity of the front and rear springs which ensures minimum pitch. At all other speeds it is more or less pronounced. It must be assumed, however, that both pitch and roll have been eliminated as far as possible, and that the suspension is designed to give maximum riding comfort at the road speed most often used, but at the same time the seat must be designed to accommodate these undesirable but inevitable phenomena.

First, consider the effect of vertical acceleration on a passenger sitting upright on an unyielding seat. The upward force impinges on the base of the spine and, unless the passenger is sitting in a bowed position, imparts a jarring shock to it which has the effect of both irritating and unbalancing him. Furthermore, the downward reaction of his own weight tends to cause him to slide off the seat.

Forward Pitch

The action of forward pitch on a double-decker differs with each position in the two saloons. Assuming that the front springs have been deflected, the body leans forward, pivoting about the rear axle so that all the passengers are tilted through a definite forward angle. This, in the early stages of the deflection, tends to hold them on their seats, but at the end of the pitch their inertia is greater than that of the bus body, and consequently they slip forward off their seats. At the front end, in the lower saloon, the distance travelled is a maximum and diminishes uniformly towards the rear axle. Behind the axle the passengers are raised, but are still subject to the same forces as those in front of it. In the upper saloon the distances travelled, and consequently the velocities attained by the passengers exceed those of the lower deck, so that if maximum comfort is to be provided seat of different design should be adopted for the two saloons.

Roll is perhaps the worst item to be accommodated. As the bus takes a corner centrifugal force causes it to roll, and the same force causes the passengers to slide along their seats.

Force of Deceleration

Acceleration holds the passengers on the seats while deceleration forces them off. The latter, being usually fiercer than the former, produces a larger force, so that there is a greater tendency to slide off the seats than to stay on them.

From the above it will be appreciated that the seat designer, to cope with all the forces set up (the principal ones being those tending to force the passengers off the seats in both a forward and sideways direction), has no easy task in front of him.

The ideal seat (if it is possible to provide an ideal seat in wood) must be of such shape that it supports the passenger in a position which demands nothing of him to retain his balance. With the back at right angles to the seat, or only slightly inclined, the occupant is precariously balanced in an uncomfortable position on the end of his spine, so that any of the numerous forces previously described easily upset him or cause him to use considerable muscular energy in maintaining his balance. The standard upholstered seat in present use has a moderately raked back which, because of its comparative softness, moulds itself to the contour of one's body and provides adequate support. A similar rake on a wooden seat, however, would prove unsatisfactory, as the timber cannot adjust itself to the passenger's back, so that he is still sitting in a fairly vertical position.

Reclining Position

The most restful position is purely reclining, and it is imperative therefore that a wooden seat, if it is to give even a minimum of comfort, should ensure that the occupant approaches as nearly as available space will allow to a reclining position without first causing the unpleasant feeling of falling which is experienced when sitting in a low chair with a too wide seat. This can only be achieved by raking the back. Furthermore, the seat back should follow the passenger's contour as nearly as possible, but, as people of different size must be accommodated, the achievement of the object is impossible. The inadequacy of the old style reversible seat back, with its single bar against which the shoulders rested, cannot be disputed, as it provided no support to the small of the back, which is imperative for a feeling of security.

If bars are still to be resorted to for economy's sake, an additional one must be provided which should be curved in shape and preferably rather low down so that the base of the spine is supported. It would be preferable, however, to provide a shorter back than usual, but of a solid shaped pattern (say, a single sheet of plywood) which would extend to the shoulder blades of the average man, a distance of approximately 15 in. In this way the support of all sizes of passengers would possibly be assured.

To cope with the forward forces tending to cause the passenger to slide off the seat, the bench position, now fitted with a cushion which fits the form even closer than the squab, must be given an additional tilt backwards. In the past many forms of seat have been evolved especially to solve this problem of forward slide, but, with entirely smooth surfaces, it has not been accomplished satisfactorily. A shaped seat canted at a steeper angle than is usual with the upholstered type should give the best results, but to ensure further comfort for the passenger a foot bar on the seat in front would prove an undoubted asset.

Width of Seat

As regards the width of the seat, the writer has conducted a number of experiments using canted boards, but has found that all widths, from extremely narrow to extremely wide (so that the knees just overhung the seat) were equally uncomfortable, and that the tendency to the disconcerting sensation of "pins and needles" was equally prevalent.

Sideways sliding due to centrifugal force must be controlled by some form of obstruction, but, as the allowable seat space per passenger of sixteen inches is already insufficient, and as this cannot be increased without running foul of the regulations regarding gangway width, it is impracticable to use arm rests to provide the obstruction. The only alternative is a shaped back and seat.

The provision of a theoretically correct wooden seat, however, would involve a great deal of expense in the machining of the specially shaped backs and benches, so unless passenger comfort is to be completely ignored and the old type straight seats provided, some other and cheaper form should be found.

Use of Canvas

Some years ago a well-known light car manufacturer used a canvas rear seat in order that it might be folded up to give a large luggage space when not in use. The seat was designed on the lines of a deck chair and could, at very little cost, be adapted for use in public service vehicles. The very nature of the canvas would be a sufficient deterrent to sliding to require no further assistance from any type of obstruction. Comfort would approach that of orthodox upholstered seats, and maintenance would consist merely of replacing and patching the worn canvas.

The powers responsible for the type of seating to be provided for the travelling public would do well to give the canvas seat their consideration before deciding on the use of expensive and unavoidably uncomfortable timber units.

The following letter is a shortened version of one which appeared in Transport World early 1947. At this time, and for four or five years afterwards wooden seated vehicles still existed, particularly in country areas where their replacement tended to take place only when an older, but more comfortably seated, vehicle was scrapped. The seats were extracted, vacuumed and prepared ready for use in lieu of the wartime efforts. It was at this point one found the floor of the bus was also rotten and the boltholes of the 'new' seats neatly corresponded with brake rods, main frames and the fuel tank, leading to some interesting 'codges'.

The foregoing article first appeared in Transport World on 11 November 1943. Looking at the seating in a Bedford OWB makes Mr. Haigh's arguments very clear. Interestingly, a number of Opel Blitz buses built in Germany to the 'Einheits' (Utility) specification had compressed cardboard doors and seats which weathered unbelievably well, while in the Angouleme district of France modified Citroen 2CV seats were used by a local operator. Presumably it was something of this kind Mr. Haigh recommended in his penultimate paragraph.

"To the Editor, Transport World,

Dear Sir,

Much has been said at conferences and in letters to the Editor, regarding the construction of chassis, body, make-up of the engine, clutch, etc., but regarding that which (we hope) completes the satisfaction of the passengers, namely the upholstery, little is said other than the type of covering used, i.e. hide, moquette, or a combination of the two. Are we to assume that such coverings form the "buffer" between the, perhaps, weary passenger and the bouncing of the vehicle? Utility seating during war years must have tried the average passenger's power of endurance to the limit, so, allowing for the necessity of temporary variations during this period due to shortage of suitable materials, might not the public vehicle passenger expect some serious research work on the question of comfortable seating? ... With pride does a manufacturer announce that a certain vehicle has done so many thousands of miles, but no mention is made of the condition of the driver's seat at the end of such a distance. Observation usually reveals jumbled springs protruding through a torn cover with a cushion to enable the driver to sit down at all ... When one compares commercial mileage with private use, can it be fair to disregard the needs of two very important members of our community - the commercial vehicle driver and the public service vehicle passenger?

Let us hope, too, that nationalisation will not leave the question unanswered nor completely ignored.

Yours faithfully,
L.I. CHESNEY

Research Department, Lintafelt Ltd."

WAR-TIME EXPEDIENCY

Governmental encouragement of home-produced fuels (in order to save shipping and conserve stocks of petrol) resulted in a variety of experiments, most of which used different forms of gas and mixtures of coal-gas and crude liquid fuel. The photograph shows one such experiment by Midland Red using a single-deck bus and a proprietory-gas unit trailer. The "Keep Off - Hot" notice on the cylinder was a very wise warning to anyone who came close, particularly cyclists who might be tempted to hang on. Petrol was used only for starting purposes but performance, little reduced on the level, was enfeebled on the hills. HA 8317 was a Short-bodied 1933 SOS, class IM6 (Improved Madam), 34 seat bus.

Guy and Sunbeam Long Service
10 YEARS OF LONG SERVICE AWARDS PRESENTED

For the first time since 1939, employees of Guy Motors, Ltd., Wolverhampton, who have qualified for long-service awards, have received gifts instead of "token" photographs. On completing 21 years of service each employee normally receives a piece of silverware - a cigarette case, a casket or a vase - but owing to difficulties during and since the war photographs, instead of articles themselves, have been given with the assurance that they would be redeemed as soon as possible. This has now been done, and at a ceremony in the works Mr. Sydney Guy, chairman and managing director, made presentations to 174 men and women. Among them was his brother and fellow-director, Mr. Ewart Guy, and two women cleaners.

In addition, a further twelve employees received certificates marking the completion of 20 years with the firm.

Mr. Guy declared: "It is only by the long and devoted service of old and good friends, and others coming along, that this company has succeeded and will succeed. That loyalty is typical of the spirit that will put this country right in the end."

Thirty-one of 61 employees of the Sunbeam Trolleybus Co., Ltd., Wolverhampton, who received from Mr. Sydney Guy certificates recording twenty or more years of service, were with the Sunbeam Motor Company when Mr. Guy was works manager before he left, more than thirty-five years ago, to found Guy Motors Ltd., which now owns the Sunbeam Trolleybus Company. Mr. Guy has now introduced into the Sunbeam concern the long-service award scheme which has been in operation in the Guy company for many years.

The Aftermath of War

Quoted from **Passenger Transport** 11 January 1950

SUNDAY COACHES

One of the facts that strikes me since petrol rationing came into force is the large number of coaches on Sunday duty compared with the number of private cars. These coaches are not on the normal excursion runs, but are engaged in taking parents to visit their children who have been evacuated to the country and seaside. No day excursions are available on the railways, and the majority of parents cannot afford the large ordinary railway fare, apart from the skeleton services available. Coach owners themselves are rationed to a comparatively small amount of petrol. An operator with four coaches who normally uses 200 gallons a week told me that his allowance had been cut down to 90 gallons.

THE TRANSPORT WORLD 12 October 1939

GOOD PROSPECTS IN THE PASSENGER FIELD

... referring to the passenger side of the industry, there is little doubt that when proper facilities have been arranged, including quick and efficient transport by sea and air, and suitable accommodation and catering, Britain will become a focus for tourists from many countries. There will probably also be a considerable influx of people who will come for business reasons, together with men on leave from our Occupation Forces. Thousands of people from our far-flung Empire will also wish to see what Britain looks like after so many years of war, and to learn, at first hand, how the Mother Country is overcoming its post-war problems. All this will mean loads for our buses, express and touring coaches, hire cars and taxis.

THE COMMERCIAL MOTOR 29 December 1944

MUNICIPAL TRANSPORT IN WAR-TIME

With the object of assisting municipal transport managers in their consideration of war-time problems, THE TRANSPORT WORLD presents a table showing the activities of the principal municipal transport undertakings on questions of services, lighting and staff.

It will be noticed that the majority of undertakings have been obliged to make reductions in either day or night services, some cases in both. Trolleybus and tramcar services, on the other hand, have for the most part not been altered.

With regard to lighting restrictions during black-outs, the methods adopted vary considerably. It would appear that all undertakings which have replied to the questionnaire have dimmed their interior lights, but not all have lacquered their windows also. The majority have preserved a form of dimmed lighting for route and destination indicators, though quite a number are shown as having no lighted indicators whatever.

Staff problems also vary. It will be seen that nearly every municipal undertaking has lost employees who have been called to the colours. The highest figure in this respect is that of Birmingham, with 1,165 employees in the Services.

So far, few undertakings have employed any women. Doncaster, Huddersfield, Manchester, Rawtenstall and West Bromwich are the only ones which are shown with women employees, Manchester having by far the largest number at 140.

[Dated 12 October 1939, the Midlands portion of this table is reproduced below, the dreariness of the vehicles was most unpleasant with one conductor likening his job to that of a hearse attendant.]

		SERVICES			LIGHTING			STAFF	
Town	No Change	Reduced Night	Reduced Day	Dimmed Interior	Blue Windows	Dim Indicators	Employees in Forces	Women Employees
BIRMINGHAM			X	X		X	1165	–
BURTON			X	X		X	7	–
CHESTER		X	X	X		X	32	–
CHESTERFIELD		X	X	X	X	X	30	–
COVENTRY		X		X	X	X	141	–
DERBY	X Trolley buses only		X Buses only	X	X	X	57	–
LEICESTER		X		X	X	X	89	–
NOTTINGHAM		X	X	X	X		200	–
WALSALL			X	X	X	X	110	–
WEST BROMWICH		X	X	X			29	2
WOLVERHAMPTON	X Trolley buses only	X	X Buses only	X			85	–

During World War II manufacture of bus chassis and bodies was strictly controlled with, reasonably enough, munitions taking priority. Initially to replace worn out and bomb damaged vehicles stocks of pre-war components were used up and then Guy Motors were authorized to produce a very basic chassis, with only primitive bodywork being made available.

In 1943 Daimler re-started production but of a new vehicle type, the CWG (Commercial, War, Gardner engine) 6, of which 100 were built before the CWA 6 became the normally available model. Although the Daimler pre-selective gearbox and fluid flywheel were still used, the engine became one of AEC manufacture, nominally 7.7 litres and developing 95 b.h.p.

The nearest bus in this photograph GLX 914, D28 in the London Transport fleet, entered service with a Duple 56 seat utility 14'6" high body in October 1944, complete with dull red paint finish and wooden seats. During repaint and overhaul in May 1948 moquette covered seats from pre-war buses were fitted but she was withdrawn in September 1953. After further service in the North East D28 was scrapped in 1958. The Cenotaph records those who died in two world wars without distinction between their regional origins.

While many women were able, quite efficiently, to fill the vacancies left by the men who had departed to serve their country in one of the armed forces, some men remained working in transport throughout World War 2 having been either forcibly detained in what was known as a "reserved occupation" or because their medical standard was below that required. In other cases they were too old at the time but while a man's age might fairly accurately be guessed, his medical condition was often not obvious and this plus the strictly applied "reserved occupation" rule could cause him to have feelings of inadequacy, if not shame. Shortly after the end of that horrendous war C.H. Morgan, writing in **Transport World**, analysed this.

Making the Job Worth While
A BUS DRIVER GIVES VALUABLE ADVICE TO RETURNING TRANSPORT MEN
By C.H. MORGAN

One hears quite a lot these days of rehabilitation schemes for men being demobbed from the Services — schemes evolved to provide a "mental adjustment on their return to civilian life."

Undoubtedly some "mental adjustment" is necessary for the men returning to the passenger transport industry. Take the attitude of the public, for instance. There is a vast difference in the way in which they regard a man in Service uniform from the way in which they regard a man in transport uniform. In addition, the returned transport man is bound to have plenty of headaches before he settles down again to the old routine and successfully handles the many pitfalls that are an inescapable part of the day's work.

But it is not only the man returning from the Services who needs to make a mental re-adjustment. We who were fortunate (or was it unfortunate?) enough to stay at our job might also examine our outlook with the object of adjusting it more closely to the value of our work to the community.

Transporting Pleasure-seekers

One particular way in which we strayed was in grumbling at having to transport pleasure-seekers, while just as loudly declaring our willingness to transport munition workers, quite forgetting that they were the same people. This grumbling culminated in a facetious use of the railway wartime slogan, "Is your journey really necessary?" a question we were always asking each other in the mere belief that ours was not.

Added to this, there began to grow a feeling almost of envy of the munition workers, not so much because of their fat wage packets, but because we became aware of the justifiable pride they took in the part that they individually were playing in the war effort. Hearing them talk as they travelled on our vehicles, we could sense how "bucked" they were at the knowledge of how the guns, tanks and planes that they had made were in use against the enemy. We could sense their satisfaction of having intimate connection with the various battles and the weapons used, and this made us acutely aware of our own lack of it.

We began to suffer from a sense of futility, and to be more convinced than ever that our journey was not really necessary. That our early risings and struggles through fog, snow and blitzes were not, after all, of much importance to the war effort. Yet even as we felt this we knew it to be wrong. We knew as well as anyone could tell us, that if we failed the whole war effort would be affected. So we became confused, and more than ever ready to see our job as one in which the day ended as it began, a round of unnecessary journeys with nothing achieved at the end.

False Viewpoint

Thus we arrived at a completely false viewpoint of our work. One, in fact, that was the exact opposite to the truth.

In common with my workmates, I had got into this false frame of mind without realising it, until one day when I overheard two girls boasting to each other of what they had made. This set me thinking, and I realised that here was much of the cause of our dissatisfaction, of our sense of futility. Everyone in a time of crisis, such as a war, takes a pride in their contribution to the common cause, yet here were we with nothing to show.

I pondered the problem for some time without getting any nearer to a solution. Then, as is so often the case, the answer came in a rather dramatic manner.

Working a late night, I was watching a crowd of pleasure-seekers being irritatingly leisurely in alighting. While they were doing so, a young fellow in marine's uniform came up to me and called out, "Thanks, pal. That was the best ride I've had for years". Vainly thinking I was the best driver on the road, I deprecated, murmuring that there were others as good as I. He laughed and said, "I'm on my way home from being a guest of the Japs. I've been in Singapore for three years". By the time I had recovered from the effective pricking of my vanity he had gone on his way.

Needless to say, on thinking it over, I became convinced that that particular journey on my part was really necessary. Moreover, at the end of the duty I felt my night's work had been worth while.

When I analysed my feelings, I realised that they were due to the certain knowledge that, in the ordinary course of my day's work, I had performed a valuable and momentous service to that young marine. Continuing in the same train of thought, I realised the possibility, indeed probability, that among the hundreds of passengers I carried each day there were some to whom the ride on my vehicle was as important as was that ride to the young marine. From this I argued that it ought to be possible to get the same sense of satisfaction each day, if only I could know for certain those to whom the journey meant so much.

I soon found the impossibility of this. However, I discovered myself picking out the passenger who appeared most worried or relieved as I drew up to the stop, and my imagination clothed their journey with all sorts of vital possibilities. Doing this, I began to find that I did get some satisfaction from my work.

Giving the matter a little further thought, I realised that here was the right way to look at our job — the right mental attitude in which to approach it.

And so we come back to the returning transport men, although it can apply to all of us. Quite a lot can be done by rehabilitation schemes, but most can be done by the man himself. If he can start off with the right mental attitude he will be well on the way to re-adjusting himself. To do this, he must begin with the realisation that his work — his "public service" — is in fact a "personal service" to his fellow citizens.

As I have pointed out, a little imagination in picking out a passenger for whom to perform this personal service helps wonderfully. Or, better still, is to acquire the certain knowledge of particular instances, like "my marine". But, whichever way he does it, so long as the transport man can retain and cultivate the idea that his work is a personal service to each of his passengers, then he can always feel that his job is "worth while". And it is only when he feels that his job is worth doing that he can develop those characteristics of politeness and cheerfulness which go to make a good public servant.

POTTERIES PRESERVATION

Potteries Motor Traction L418 was rather a vehicle rarity for the Midlands, but was acquired with the business of Mainwaring Brothers, Bignall End, in June 1951. Built 1949, SRE 563 is a Crossley DD 42/7 with a Crossley built 53 seat lowbridge body. She was withdrawn from service in 1960.

To single out any one restoration is difficult, but Christopher Meir, the owner and restorer of TVT 128G has been careful to keep a photographic record: the following is a shortened version of his history of this PMT vehicle.

TVT 128G was new to PMT Milton depot on 1.1.69 and was one of a batch specially built for PMT for service around the Cheadle area. The Leyland Leopard PSU4A/4R chassis was fitted with a low crawler gear for steep gradients while its 43 seat Marshall body was built of aluminium alloy in order to reduce the vehicles overall weight. 128 remained at Milton until the depot's closure in 1980, having the honour of being the last bus to return to the depot.

After withdrawal from passenger service 128 was transferred to the PMT Driver Training School, painted in an all over yellow livery and carried the fleet number T3.

Final withdrawal came in 1986 when the vehicle was delicensed and put into store at PMT's Clough Street depot, where it remained until it was purchased for preservation in April 1990.

TVT 128G was later moved to its new home in Stone where it is currently undergoing full restoration to as near original condition as possible.

THUMBNAIL SKETCH 3 - BROWN'S BLUES, MARKFIELD

Brown's Blues started operating in the classic manner when Mr. L.D. Brown purchased a vehicle which served as coal lorry-cum-bus, undergoing a body conversion each change of use. In 1927 passenger demand in this part of Leicestershire having risen, the first real coach was purchased.

An attempt to summarize the later growth of Brown's Blues is difficult - what other company in its 40 years of independant life could have operated around 40 different models of chassis from 11 makers and utilized the products of 17 bodybuilders?

Although a number of smaller concerns were purchased, Brown's Blues kept within their own patch bounded by Leicester, Hinckley and Ashby for bus operations, although their coaches in a rather superb livery of deep blue relieved by primrose and red were from time to time to be found at most popular destinations.

The end came on the night of 15 March 1963, when Midland Red absorbed the business.

GUT 400 AEC Regal IV/Yeates C41C in service 1951 - 9/60

EJU 260 Dennis Lancet III/Yeates C35F in service 1947 - 2/63

ENR 294 Bedford OB/Duple Vista C29F in service 1954 - 3/58

FAY 517 Daimler CVD6/Brush H56R in service 1949 - 11/61

HLW 166 AEC Regent III/Park Royal H56R in service 1958 - 3/63

GUT 398 Leyland Royal Tiger/Duple Roadmaster C41F in service 1951 - 3/63

KUG 6 Leyland PSI/Wilks & Meade FC 33F in service 1958 - 9/60

FROM WEST MIDLANDS TO EVERYWHERE

As a consequence of de-regulation in October 1986, several of Britain's larger bus operating undertakings suddenly found that due to service reorganisations and the tendering process, they had a large quantity of vehicles which had become surplus to requirements. In particular, Greater Manchester PTE and West Midlands PTE both had double deckers for which they had no further use and as a result, offered these for sale to other operators. Many of the buses concerned were little over ten years of age and as such were regarded as good purchases by large and small operators alike and quickly they became familiar sights in almost every corner of Britain. Perhaps because Greater Manchester employed a major dealer to dispose of its surplus buses, a large quantity of these found their way into the fleets of major operators who, at that time, were in need of additional vehicles with which to operate their expanded networks. West Midlands' double deckers however, with one or two exceptions, appear to have not attracted as much interest from the larger companies but instead appealed more to the smaller independant concerns. In a bid to standardise its fleet during its period of reduction, West Midlands took the decision to first rid itself of its Ailsa Volvos and Bristol VRTs before starting to make any sizeable inroads into its large quantity of Daimler Fleetlines and to this end began to offer its double deckers of the first two mentioned types for sale towards the end of 1986.

Amongst the first operators to take an interest in West Midlands' surplus double deckers was London Buses who had just purchased a number of Van Hool McArdle-bodied Ailsa Volvos from South Yorkshire PTE. This was somewhat surprising as prior to this, London had operated only three Ailsa Volvos (which had been taken into stock for evaluation purposes in 1984) and which had originally been operated from Stockwell garage. After a brief period in store during 1986, this trio was moved to Potters Bar garage where they were ultimately joined by the South Yorkshire examples at the beginning of 1987. Obviously finding these buses suitable for their requirements, London Buses entered into negotiations with West Midlands Travel during the early summer of 1987 with the result that 50 Alexander-bodied Ailsa Volvos were purchased from that undertaking. These began to arrive in their new home during the latter part of 1987 and their delivery was completed early in the following year. 24 were painted in London's traditional red livery for use from Potters Bar garage whilst the remaining 26 were given a livery incorporating a large area of cream and, adorned with Harrow Bus fleet names, were placed in service from Harrow Weald garage.

Meanwhile, former NBC subsidiary Wilts & Dorset, who after de-regulation were seeking a number of additional double deckers to meet their increased requirements, purchased 15 MCCW-bodied Bristol VRTs from West Midlands during 1987. Painted into their new owner's attractive red, white & black livery, they were quickly pressed into service on a wide variety of duties and soon proved to be popular with drivers and passengers alike. They were, however, withdrawn by early 1992.

North of the border, Western Scottish were also in need of additional double deckers with which to enhance their fleet and upon learning of the availability of buses from the West Midlands, in the early summer of 1987 purchased 11 Park Royal-bodied Daimler Fleetlines from that undertaking. These complemented Western's own fleet of buses of this chassis type and after receiving their new owner's white, black & grey livery, they were placed in service from Ayr depot before some were later dispersed to other of the company's operating centres.

As with those sold to Wilts & Dorset, the Western Scottish Fleetlines only had a five year extension to their lives.

Another operator to purchase former West Midlands Fleetlines was Derby City Transport who in 1987 acquired 2 Park Royal examples as replacements for a pair of their own Fleetlines which had suffered serious accident damage. More recently, Maidstone Boro'Line added 4 West Midlands FE30AGR-type Fleetlines to its double deck collection when in 1990 it purchased 1 Park Royal-bodied and 3 MCCW-bodied examples, all of which were immediately placed in service in an all-over white livery.

These then were the only former West Midlands double deckers to find new homes with large operators, although they were by no means the only deckers from this undertaking to find a further lease of revenue-earning life. Throughout 1987, West Midlands' surplus Bristol VRTs and Fleetlines were given new homes by independant operators all over Britain from as far afield as the south coast to north of the Scottish border and wherever enthusiasts travelled, one or more of these buses was almost certain to be seen.

In the south, apart from those already mentioned, 2 Fleetlines were added to the fleet of Hampshire independant, Jones of Oakley in 1987 and not too far away, Pike of Smannell purchased a pair of MCCW-bodied Bristol VRTs in 1988 and Brixham Coaches took a pair of Fleetlines. In addition to these, 2 of the Ailsa Volvos previously operated by London Buses were later acquired by Collins, Castle Goring whose fleet had previously contained an ex. West Midlands Fleetline.

Whilst the south west failed to provide a home for many exiles from the West Midlands except for one or two employed in non-PSV roles, in South Wales these vehicles could be seen in service with several operators. Amongst these was Davies, Pencader; Coastal Continental of Barry; Morris, Pencoed and Thomas of Porth who between them operated 10 Fleetlines, and Lewis, Carregefan; Williams, Porthcawl and Silcox, Pembroke Dock whose fleets contain 4 Bristol VRTs. Several of these have seen service with other operators in between leaving their West Midland home and arriving in South Wales. North Wales also boasts a number of ex West Midlands buses in the form of Fleetlines with Lloyd, Baggilt and Whiteways of Waunfawr while Clynog & Trefor and Wright, Wrexham both had Bristol VRTs from this source, as did Prestatyn Coachways.

Moving northwards to Merseyside and Lancashire, this part of the country has seen West Midlands vehicles operating in a variety of fleets with Cherry, Bootle, Loftys of Ellesmere Port, Powny, Aintree and Forrest (Blue Triangle), Bootle having one Fleetline each and Arena, Liverpool having 10 buses of this type which all ultimately passed to Amberline. Reilley, Bootle broke the mould by operating a Bristol VRT acquired via Prestatyn Coachways while Hill of Congleton, Cheshire also had a pair of VRTs, one of which had previously been operated by Brixham Coaches in Devon. Although the independant operators in Greater Manchester have tended to favour former Manchester double deckers, Stotts of Oldham purchased four Fleetlines from West Midlands while Stuarts of Hyde bought three. Further north, Mercers of Longridge operated a number of Fleetlines until the company's stage carriage operations were taken over by Ribble and Lonsdale Coaches of Heysham also ran both Fleetlines and a Bristol VRT for a while. This was not surprising as, in their dealer capacity, Lonsdale handled several West Midlands vehicles including those added to their own operational fleet. After the company's sale to Lancaster City Transport, the VRT continued to run for its new owner for a short while until it was eventually replaced by an ex.Highland Scottish Leyland Atlantean.

Crossing the Pennines, Yorkshire has proved to be the largest operational area outside London for former West Midlands vehicles with Fleetlines, Ailsa Volvos and MCW Metrobuses all to be seen. Without doubt, the largest operator of vehicles from England's second city is Andrews of Sheffield who during 1990/1 purchased no fewer than 18 Fleetlines for operation on new services in and around their steel city home. Painted in a bright yellow & blue livery, these look immaculate as they frequently pass through Sheffield city centre.

Further Fleetlines followed in 1992 to bolster the WMPTE presence in this fleet.

Amberley Coaches of Pudsey, near Leeds have also given a new lease of life to several Fleetlines and indeed have replaced some of their earlier examples with newer buses from the same source. White Rose of Castleford also operated a solitary ex.West Midlands Fleetline whereas Leon of Finningley bought four. Possibly as a result of a number of London's ex West Midlands Ailsa Volvos being sold to south Yorkshire dealer and breaker, Wombwell Diesels, several of these buses gained a stay of execution when they were snapped up by other operators who realised that they still had more than a little life left in them. Black Prince of Morley, on the outskirts of Leeds, whose fleet already contained a couple of the well-travelled former South Yorkshire Van Hool McArdle-bodied Ailsas added to these three of the ex.London examples in 1989/90 whilst on the other side of the city, Four Seasons of Woodlesford purchased two in 1991. Further West Midlands buses to find favour in Yorkshire were a pair of Fleetlines acquired by the now defunct Driffield & District and a trio of similar buses which were given a new home by Primrose Valley of Filey. These later passed to the EYMS Group upon their acquisition of the Primrose Valley business and these were later joined by one of the ex.Driffield & District examples. They were, however, withdrawn by the EYMS Group in 1992.

Further south, Skill, Nottingham bought four of the ex.London/West Midlands Ailsas, later reselling two of them to Tame Valley, Birmingham, curiously, these were taken out of service in 1992, when the company was, effectively, taken over by WMPTE; a rare example of a circle squared!

Kettlewell, Retford took two and Scutt of Owston Ferry acquired one while another Owston Ferry independant, Bannister instead chose a Fleetline, as did Kime, Folkingham and Simmonds (Reliance), Great Gonerby. In Leicestershire, the British Shoe Corporation purchased a couple of Fleetlines to add to their large double deck staff bus fleet, Astill & Jordan acquired a similar bus and Smith of Market Harborough bought a Park Royal-bodied example in 1987. The largest fleet of former West Midlands Fleetlines in this area was, however, built up by G.K. Kinch of Barrow on Soar who used them on his expanding network of stage carriage services upon which they looked distinctive in their yellow & blue livery. Three of these buses, together with the Astill & Jordan example were later sold to Midland Fox in 1988/9 respectively. Charter Coaches of Gt. Oakley also ran a number of ex. West Midlands Bristol VRTs until the company's demise, these then passing to Eastern National for eventual resale.

Surprisingly few West Midlands double deckers have found new owners in the north-east however, although Northumbria's subsidiaries Curtis Coaches and Moor Dale Coaches' fleets each contain two Park Royal-bodied Fleetlines whereas north of the border almost twenty West Midlands deckers are currently in service. Amongst these are Fleetlines with MacConacher, Ballachulish (2), Crawford, Neilston (1), Gilchrist, East Kilbride (2), Keenan, Coalhall (2), Liddell, Auchinleck (4), Silver Fox, Paisley (1), Earnside, Glenfarg (1), Smith, Patna (1), A1 Service (McKinnon, Kilmarnock) (2) and Moffat & Williamson, Gauldry (1). A1 are one of the operators who seem to have the knack of making second hand vehicles work well and have strengthened their fleet during 1992. Marbill of Beith took a Fleetline in 1992, and in addition, Moffat & Williamson also have 2 MCCW-bodied VRTs while Marshall, Balieston also have a solitary example of this type, as does Earnside, Glenfarg.

A batch of wandering ex-WMPTE Fleetlines were those operated by Wilson of Greenock (Pride of the Clyde) as most are now in the hands of new owners south of the border.

The only ex.West Midlands Ailsa exiled from London was purchased by McColl, Bowling during 1991.

More recently, West Midlands Travel has begun to withdraw some of its large fleet of MCW Metrobuses, and some of these have already attracted interest from other operators. As might have been expected, Stevensons of Spath - whose fleet is well known for its interesting second-hand vehicles - were amongst the first to offer these Metrobuses a new home and in addition, Yorkshire Traction purchased five in 1990, painting these in their County Motors livery of cream and two tone blue. No doubt in the months and years ahead, more of these buses will start to appear in other fleets and may in time warrant an article on their movements.

Mention should also be made of the large number of ex.WMPTE minibuses which have found new owners in various parts of Britain. Typically, West Yorkshire PTE had a quantity of Freight Rover Sherpas - 10 in all - with Carlyle bodywork which they acquired in 1987. Although most only had a three year life extension a few survived until the autumn of 1991 in the Leeds and Halifax areas, by then in Yorkshire Rider livery.

As stated at the start of this brief article, although not as numerous as the former Greater Mancheser double deckers which have found new homes, those from West Midlands Travel have proved to be far from redundant and have added much of interest to fleets all round Britain. Numerous more have found a new lease of life in non-PSV roles such as staff buses, playbuses and promotional vehicles and undoubtedly many will see a good few years more life before being eventually despatched to the breakers yards. Despite all the foregoing however, only a comparatively small proportion of the buses discarded by this large midlands undertaking continue to survive and, following the large number which have already entered the yards of the Carlton breakers to be quickly reduced to piles of tangled metal since 1986 have been several of those which have enjoyed a brief reprieve. Those of Western Scottish, Wilts & Dorset and Derby along with over half of the Ailsa Volvos purchased by London Buses have also now run their final journeys whilst others are now with their third or fourth owners.

[The foregoing first appeared in Bus Fayre, January 1992, and was updated to June that year by K.A. Jenkinson. It is reproduced by permission of Autobus Review Publications. Readers' attention is drawn to the unique format of Bus Fayre, as it alternates monthly between a general bus magazine and an in-depth study of specific chassis or body manufacturers.]

Ex WMPTE Gardner engined Fleetline 6412, NOC 412R, resprayed in 'Dealer White' stands in Thomas's (ex National Welsh) garage at Porth, 11 January 1992.

One of Liddell of Auchinleck's former West Midland Fleetlines, EOF 297L rests at its owner's depot on a Sunday in April 1992. It is normally used on contract and schools services.

One of a pair of ex West Midland Fleetlines owned by McKinnon, Kilmarnock (A1 service), KON 336P stands at its owner's depot alongside NOC 380R in April 1992.

Ex.WMPTE fleet number 4593, this Daimler Fleetline has been re-badged as a Leyland DAF, as well as being partially gutted by its new owners.

JOV 746P looks incredibly smart in its new Trailways livery, belying its 16 years of hard city life.

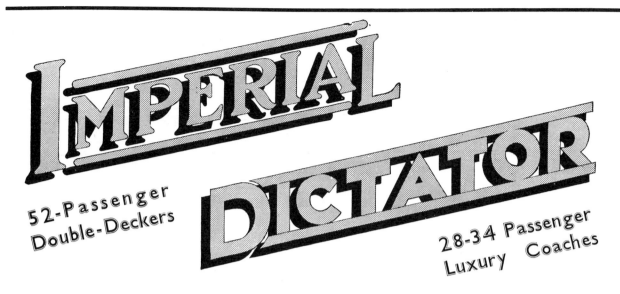

52-Passenger Double-Deckers

28-34 Passenger Luxury Coaches

The Passenger Vehicles with the WHEEL-OUT ENGINE

Built from the Operator's point of view, to enable repairs and replacements to be carried out with a minimum of time and labour. Efficiency and economy is ensured by the patented high camshaft and overhead valves.

Orders received from such important users as Birmingham Corporation, Edinburgh Corporation and the East Kent Road Car Co., Ltd., prove the definite superiority of Morris-Commercial passenger vehicles.

Full particulars, catalogues, etc., on request

MORRIS-COMMERCIAL
THE LARGEST BRITISH MANUFACTURERS SPECIALISING EXCLUSIVELY IN COMMERCIAL VEHICLES

MORRIS COMMERCIAL CARS LTD — ADDERLEY PARK — BIRMINGHAM 8

OMNIBUS AND COACH SECTION

In this series of books it is our intention to take any one year in the life of motorized public carriages within the Midlands (and outside where events were later to affect the Midlands) and draw from contemporary material some of the flavour; in Cordon Bleu terms we shall shake a drop or two of essence from the bottle of vehicular tincture - that mixture of oil, fuel, wood, moquette, leather, linoleum, paint, people and paperwork that is a public service vehicle.

The following extracts are culled from the "Omnibus & Coach Section" of the Tramway and Railway World published in the year of Our Lord 1929.

NECESSARY IMPROVEMENTS TO VEHICLES.

Recently there has been a storm of complaints in daily newspapers concerning the operating methods of some coaching undertakings. In many cases the operators have been much to blame, but it may be said truly that they are only a small proportion of the large number of coach owners in this country.

Even in the larger and better regulated concerns, however, although they are not guilty of the shortcomings urged against some of the lesser firms, there are things that might be done to secure the greater comfort of passengers.

It is unreasonable to expect that a coach two years old, will be quite as comfortable as a new vehicle. It will certainly be somewhat shabby. There is no reason why, at a small cost, a little leather reviver should not be used which would give the interior a brighter appearance. Seat squabs could be re-sprung. Grab handles, rug rails and other fittings become loose through the continued vibration, but in some instances no attempt is made to tighten them. These are small details that often irritate passengers, besides adding to the noise. It is recognised that there is another side to the question. It is not worth spending too much on a two-year-old coach; one-half of its useful life is over. A relatively small amount spent at the right time would, however, keep the vehicle in a better condition and prevent unfair comparison with the latest coach on the road.

A SOURCE OF GREAT DANGER.

The part of an omnibus or coach chassis that is most likely to give rise to accidents is the braking. Nowadays mishaps due to defective brakes are of rare occurrence, but complete immunity is not to be looked for unless the vehicle is provided with an air-brake, in addition to the customary retarding of the wheels and the transmission.

The need of daily attention to adjustment was strikingly brought to our attention recently on a coach journey of over two hundred miles in a vehicle that had come from one of the leading makers only a few weeks prior to this particular trip. An hour out from the starting place, on a village road that was very little wider than was needed for two lines of traffic, the driver proceeding at reduced speed was suddenly confronted with an omnibus imprudently pulling over to the centre of the road to avoid a stationary vehicle. Contact between coach and omnibus at about their central point seemed to be inevitable, although the coach-driver had applied his brakes. The coach came to rest when only an inch or so separated the two vehicles, and it was evident that the former had travelled further than the driver had expected it to do. Adjustment of the brakes was effected in less than a minute through an opening in the floor behind the second row of seats. During the hour's run there had been trouble-free application of the brakes on a number of occasions due to traffic ahead, a fact that leads to the surmise that the slackness met with in the emergency had a sudden origin and could not have been detected by tests which should be applied immediately after leaving the starting place.

SAFETY ON THE ROADS.

A suggestion is that drivers of road vehicles should be examined, as railway drivers are, as to their physical fitness and their knowledge of traffic rules and regulations. It is possible that legislation to this effect would result in hardship to many individuals. This, while unfortunate, should not be allowed to prevent the adoption of a general measure that would tend to diminish even slightly the risks attendant on travel by road.

HAND SIGNALS.

Uniformity in hand signals will always be an ideal extremely difficult to obtain, for the simple reason that each individual has his or her own views of what constitutes a signal. As, however, the misinterpretation of a signal may easily cause a serious accident, too much care cannot be observed in its use. All modern omnibuses and coaches are so constructed that very definite signals can be made which will clearly indicate the intention of the drivers, and as these are generally well trained and experienced men, little difficulty is experienced from this class of road traffic. But with the great increase in the number of small cars with fixed side screens, a very serious element of danger has been introduced. In these cars not only is the provision for signalling inadequate or non-existent for those who really wish to conform to the rules of the road, but, by the absence of facilities, distinct encouragement is given to the careless driver to ignore signals entirely. It is a common sight when approaching cross roads to see the driver of a private car frantically waving his hand through an opening in the roof without giving any definite indication of the direction the car will take. The system of hand signals should, as indeed the Royal Commission on Transport recommends, be made as uniform as possible throughout the British Isles.

OBSOLETE PASSENGER VEHICLES.

A matter that urgently calls for drastic action is the risk of accident and consequent danger to the public, from the running of obsolete vehicles in passenger services. In the days of the horse-drawn vehicle it was the practice for persons able to acquire a wagonette and an animal to draw it, to take advantage of a rush of traffic, and as anything on wheels would be granted the necessary licence, many sorry looking combinations found their way into public service. The place of these old rattle-traps has, unfortunately, been taken by a type of motor vehicle which is almost the exact counterpart of its predecessor of the horse days, and this to a much greater extent than is generally known.

THE QUESTION OF SPEED.

In his evidence before the Royal Commission on Transport Colonel Pickard said that excessive speed, having regard to all the circumstances, is the greatest cause of accidents presumed to be avoidable; but, he added, excessive speed may have been only five miles per hour. To a direct question he replied that in his opinion a rigid speed limit was not a good thing from the public point of view. With the increase in the size of passenger-carrying vehicles, with covered double-deck omnibuses running on country routes and the rapidly increasing service of long-distance coaches, it is a difficult question what form of control should be placed upon this class of traffic for the safety of the public. The present 20 miles per hour limit for pneumatic tyred vehicles might not exist for all the attention that is paid to it by the majority of drivers. Nevertheless it is the law, and all who exceed it are therefore law breakers. Clearly matters cannot remain as they are; but whether a maximum speed limit be placed upon certain types of vehicles, or speeds of varying degree permitted on certain routes, the human element will still remain, and greater care in the selection of training of the drivers of public service vehicles will have to be exercised in the future if accidents are to be avoided, especially where competing operators are concerned.

RESPONSIBILITY OF OWNERSHIP.

From time to time the question of responsibility for lapses on the part of drivers of public service vehicles is raised in the police and civil courts. As the nature of the lapse varies to a considerable degree, it is difficult for any authority to lay down a hard and fast rule. For instance, in a case of plying for hire with an un-licensed vehicle, much depends upon proof of the instructions given to the driver, as it is an easy matter for an employee to do many things without the knowledge or consent of the proprietor. Then on the question of speed - about which we may expect to hear of many prosecutions if the chief constables of the provinces act upon the intention which has been expressed in various quarters - the time-tables and conditions of the service will be a prominent factor in determining the responsibility. While the personal offences of the driver, in relation to the passengers and to the local authority, must always be a matter of his own liability, there have been cases in which a driver has been unfairly penalised on account of faulty vehicles, the responsibility for which appears to rest entirely upon the owner. One serious offence of this kind relates to inefficient brakes. It is an open question whether large vehicles should be permitted to use the narrow roads adjacent to many charming resorts favoured by trippers. There can, however, be no question that to permit vehicles to be thus used without ample brakes is an offence meriting severe punishment. That this view is held to some extent by the Aberystwyth magistrates was recently indicated. The owner and driver of a coach were summoned for using, and allowing the vehicle to be used, with inadequate brakes, and in ordering the owner to pay £15 the chairman of the bench expressed the justices' determination to hold the owners responsible for the offences of the drivers whenever possible.

SPEED GOVERNORS IMPRACTICABLE.

Lord Apsley, in the House of Commons, asked the Minister of Transport to consider making a regulation that heavy vehicles should be fitted with a governor so that the engines should not exceed a maximum speed of 30 M.P.H. Col. Ashley replied that the question was considered by the Departmental Committee on the Taxation and Regulation of Road Vehicles. The Committee were of opinion that the proposal was impracticable, on account of the engineering difficulties, the expense, and of the facility with which such a mechanical arrangement could be altered.

THE PETROL SITUATION.

On January 5, the Power Petrol Company announced a reduction in the price of petrol for commercial vehicles from 1s. 4¼d. to 1s. 2d. An earlier announcement by the Motor Agents' Association credited the company with having agreed to the petrol combine's arrangements regarding dealers' profits and pump sites. The association further stated that the company were setting aside 5 per cent. in respect of compensation to expropriated Russian shareholders. At a meeting of the Anglo-American Oil Company on December 27 it was stated that owing to continued depression no interim divident would be paid.

PROPOSED CORPS OF TRAFFIC POLICE.

Replying to a question in the House of Commons, Sir W. Joynson-Hicks, Home Secretary, stated that he had for some time had under consideration the question of holding a conference on the subject of instituting a corps of traffic police. The County Councils' Association and the Association of Municipal Corporations had now, at his invitation, nominated representatives to a conference, to be held early in 1929. It will be remembered that numerous county councils have lately pressed the proposal on the attention of the Home Office with a view to the cost of traffic regulation being borne by the Road Fund.

CAUSES OF CAR BREAKDOWNS.

The Royal A.C. have issued their annual statistics showing the causes of car breakdowns dealt with under their "get-you-home" service. An analysis is subjoined, the figures in parentheses being the percentages for 1927: Ignition 22.7 (21.0); carburation 2.6 (2.6); valve mechanism, camshaft, lay shaft, secondary and timing gear 0.7 (0.4); valves 1.4 (1.4); lubrication 2.5 (2.7); water circulation 2.2 (1.1); crank-shafts 0.5 (0.5); starting 1.0 (0.9); clutch 5.1 (4.9); gear-box 2.6 (3.2); couplings, universal joints and propeller shafts 4.1 (4.9); brakes 0.1 (0.2); back axle shafts 13.8 (13.6); differential 0.7 (1.4); bevels and worms 1.0 (0.7); front axle and steering 3.3 (3.1); road wheels and suspension 4.2 (4.1); and lighting failures 2.0 (2.2).

BROKEN GLASS.

Not the least of the risks attaching to travelling by road is the possibility, in even a slight accident, of severe injuries to passengers due to the fragility of ordinary glass. There have been many cases in which passengers in public vehicles have been cut by broken glass. It is true that in some instances the injuries are not in themselves severe, but there is always the chance that such wounds may become septic, since it is impossible to keep the windows of public vehicles sterilised. The risk may not be great, but it is one which the public should not be expected to run. Should the processes adopted prove satisfactory there is no reason why the operators of public vehicles should not decide on new vehicles being fitted with unsplinterable glass. Even with lower prices the cost will be greater than that of ordinary glass. It would seem reasonable, however, if third party insurance becomes compulsory, that insurance companies should offer reduced premiums for vehicles equipped with safety glass windows. The use of such glass is, of course, not entirely a matter of first cost. The technical properties of several makes of unsplinterable glass are not yet fully known. Only extended experience can show whether these makes will remain reasonably colourless, or whether under prolonged vibration the adhesion of the outer glass layers to the inner material will be maintained. Thorough tests under working conditions are required to determine the best makes to adopt, and the sooner the tests are made the better for the public, the operators and the safety glass industry itself.

NEW SERVICES FROM LONDON.

Thomas's Saloon Coaches, of Eccleston Street, London, are running twice daily to Swansea, and the Great Western Express Co. has increased its London-Cardiff service to four journeys each way daily. The latter company has opened an office at Great Newport Street, W.C.2, where a waiting room is also available. Further competition on the London-Cardiff route is provided by a service three times a day in each direction by the Westcliff Motor Services, Southend, the London terminus and booking office being the office of Road Travel Bookings at Bush House, Aldwych.

LEASED SERVICE AT WORCESTER.

In the early part of last summer Worcester Corporation accepted the offer of the Midland Red Omnibus Co. for the exclusive use of the city's right to operate omnibuses. Ald. W.J. Hill, chairman of the Traffic Committee, at a recent meeting of the Council, explained that owing to circumstances over which they had no control the Council had to incur an outlay of £75,000 for purchasing the undertaking of the Worcester Electric Traction Co. and abandoning the tramway service. The loan had to be repaid within twenty years, and the Minister of Transport stipulated that whatever profits were made during that period from an omnibus service should go to repayment of the loan. The sale of the tramway assets was estimated to produce £5,000, reducing the capital indebtedness to £70,000. Loan charges were £6,000 per annum, and to avoid any cost falling on the rates they had to receive £500 a month. This amount was at present being more than met, as for June, July, August and September, the receipts were £3,400. In the last complete year of the tramway company the mileage run was 280,000, but for the first four months of the omnibus services the mileage was 190,000 in the city only, and the number of passengers was 2,351,039. The population of the city had been carried 46 times in that period.

RED REAR LIGHT QUESTION.

The justices recently heard a summons against an omnibus driver for failing to have a red rear light on his vehicle. The light had become extinguished, but a solicitor, instructed by the C.M.U.A., contended that the light was not necessary as there was a red glass window at the rear of the omnibus and an index number illuminated by the lights in the vehicle. The case was dismissed on payment of costs.

MINISTRY OF TRANSPORT AND SLIPPERY ROADS.

Road authorities have received a circular letter from the Ministry of Transport, stating that as the result of investigations, the Minister has arranged for the publication, through the British Engineering Standards Association, of seven specifications representing the best practice of the day for the prevention of skidding on roads, which should henceforth be adopted by local authorities.

WAR OFFICE SUBSIDY VEHICLES.

The Secretary of State for War has informed the House of Commons that the number of vehicles to be subsidised under the present scheme was limited to 1,000, and that number was at present enrolled. During 1928 the subsidy of £40 a year for three years was paid for 484 vehicles in replacement of a similar number which had ceased to be eligible. The total payments from January 1924, to January, 1939, amounted to £147,000.

NEW OMNIBUSES FOR BIRMINGHAM.

In view of the increasing traffic on omnibus routes Birmingham Tramways and Omnibus Committee have ordered a further fifteen omnibuses. These will be of the latest type, having six-cylinder engines. There will be a single step to the platform and standard seating on the top.

EDINBURGH-GLASGOW ROAD.

In reply to a question concerning the progress of the Edinburgh-Glasgow Road, the Minister of Transport has stated that of the total length of 39½ miles, 27 miles were completed, but of this length 11 miles could not be brought into use until certain bridges were constructed. Five miles were under construction and the remaining 7½ miles of reconstruction of an existing road would shortly be put in hand. The total estimated cost was about £2,200,000, of which amount approximately £1,100,000 had been expended. The work on the 7½ miles referred to had been intentionally delayed until the cost of the remaining sections could be more closely determined.

LONDON-BIRMINGHAM.

Early in January a "Midland Express" coach, owned by Senior Service Coaches, Upper Street, Islington, N., began a daily non-stop service between London and Birmingham. The starting point is from York Road, near King's Cross Station. The coach is a Gilford 26-seater, and the fares are 9s. single and 15s. return (28 days), as against 13s.11d. and 27s.10d. by rail.

LONDON TO EDINBURGH IN A DAY.

To meet the traffic offering since the discontinuance of the two-day journey from London to Edinburgh in the tourist season, E. Thomson, of 16, Princes Street, Edinburgh, is running a service twice weekly from 60, Haymarket, to and from the Scottish capital. The journey is made in a day, at a fare of 30s. single and 50s. return.

AN URBAN COUNCIL SEEKING POWERS.

A Bill is to be promoted by Oldbury Urban Council for powers to run omnibuses.

RAIL-ROAD COMBINED SERVICE.

Great success has attended the combined road-rail service of the Great Western Railway between Cheltenham-Oxford-Reading and Paddington. There are five daily and three Sunday services, each way, between Cheltenham, Reading and Paddington, and in addition four daily services each way between Burford, Reading and Paddington. A late Sunday night service has been provided in both directions between Cheltenham, Reading and Paddington. Luxury cars operate the road journey between Oxford and Cheltenham.

LICENSING CONDITIONS AT STOKE-ON-TRENT.

The Watch Committee recommended the renewal of omnibus licences for the ensuing year subject to the following conditions:- The licensee to effect an insurance or make adequate financial provision for meeting any liability for injury or damage occasioned by his vehicle; maintain a regular service to a time-table to be deposited with the chief constable, such time-table not to be varied except by seven days' notice; an omnibus to be of a design and equipment suitable for its route; the Corporation to have the right to appeal to the Minister of Transport against the maximum fares charged; the whole of the route to be covered each journey.

BRITISH MOTOR COACH SERVICES ASSOCIATION.

Presiding at a meeting of this new Association, held at the Charing Cross Hotel, London, Mr. G. Nowell (Great Western Express) stated that its objects would be to represent coach proprietors in all matters, especially with regard to proposed legislation. All operators of long-distance vehicles would be welcomed, and it was intended to press for co-ordination in licensing and the extension of the 20 M.P.H. speed limit. Efforts would be made to organise a co-ordinated system of road transport all over Great Britain, with tickets to any part of the country. The entrance fee was fixed at £5 per firm, and the annual subscription at £2 2s. Mr. K. Pocklington, 12, Shepherd's Bush Green, London, W., is acting as hon. secretary.

BIRMINGHAM TO WESTON-SUPER-MARE.

Ten licences have been granted by Cheltenham Watch Committee to the Birmingham and Midland Co. for new omnibuses running between Birmingham and Weston-super-Mare.

LICENCE ANOMALIES.

An omnibus-owning firm and a driver were fined nominal sums at Leicester for not having the local licence for an omnibus which had been sent from another town to Leicester to relieve the overcrowding of another vehicle which was locally licensed. In a summons at Derby against the North-Western Omnibus Company for permitting an omnibus to be used in that borough without a licence, it was explained that the vehicle was a substitute for one which was under repair, and which was duly licensed. The substitute did not, however, meet with certain requirements imposed at Derby, where the regulations differed from those at Manchester and Nottingham. A fine of 10s. was inflicted, but a summons against the driver was dismissed on payment of costs. It is expected that the Road Traffic Bill will remedy difficulties such as the foregoing by providing for a uniform licence.

A REMARKABLE FUEL.
Under the title of "Blupetrol" a new fuel, for which sweeping claims are made, has been advertised by the Blue Bird Petrol Companies, 25-31 Moorgate, London. The producers guarantee that it will increase mileage and pulling power by 15 to 20 per cent., ensure easy starting, prevent carbon deposits and actually remove existing carbon, eliminate knocking and pinking, have no harmful effects on any part of the engine, reduce running costs, and prolong the life of the engine.

THE SPEED QUESTION.
Acting on Home Office instructions, numerous prosecutions have taken place of coach drivers and owners for speeds in excess of the new limit of 20 M.P.H. in the case of heavy vehicles running on air tyres. Bath Corporation, on January 1, decided not to licence services having an average speed, according to time-table, exceeding 17 M.P.H.

A NEW G.W.R. SERVICE.
On August 1 the Great Western Railway started an omnibus service between Wolverhampton and Aberystwyth. The vehicles leave Wolverhampton at 11.48 a.m. and Aberystwyth at 10.30 a.m. daily, including Sundays. The fares are 11s. 6d. single and 21s. return, the corresponding third-class railway fares being 14s. and 28s. respectively.

BIRMINGHAM-WESTON-SUPER-MARE.
Six licences have been granted by Cheltenham Watch Committee to the Birmingham and Midland Omnibus Company to increase its service between Birmingham and Weston-super-Mare, via Gloucester and Bristol, by an alternative service via Cheltenham and Bath.

IN PLACE OF A TRAIN SERVICE.
The London, Midland & Scottish Railway on August 3 started an omnibus service in conjunction with the Mansfield & District Tramways, Ltd. The route is between Mansfield and district and connects with the railway stations at those places and on certain routes with the stations at Southwell. The passenger train service between Mansfield and Southwell has been withdrawn, and Blidworth, Rainworth, Farnfield and Kirlington stations have been closed for passenger traffic.

ATTITUDE OF BIRMINGHAM WATCH COMMITTEE.
While refusing all new applications for licences to ply for hire in the city, on the ground that traffic is congested, the Birmingham Watch Committee are informing long-distance operators that they must obtain premises from which to run, thus complying with Section 45 of the Town Police Clauses Act, 1845.

THE PNEUMATIC TYRE SUPREME.
At December 31 last the Birmingham and Midland Omnibus Company had 600 omnibuses and sixty coaches all on pneumatic tyres. The Midland "Red" omnibuses operate one thousand services, and have a route mileage of over 6,650. All the company's vehicles have now been converted to pneumatic running.

TRAFFIC GROWTH IN THE MIDLANDS.
The census of traffic taken in August last on main roads of Warwickshire shows that the average increase in weight in tons per day was 45 per cent. over the year 1925, 160 per cent. over 1922, and 911 per cent. over 1913.

MIDLAND "RED" PLANS.
The Birmingham & Midland Omnibus Co. intend to start daily services from Birmingham to London, Manchester, Liverpool and Bournemouth.

AN INVITATION TO THE RAILWAY COMPANIES.
The Transport Committee of Chesterfield Corporation have invited the railway companies to investigate the financial side of the undertaking and to present offers for (1) the routes classified in two groups, and (2) for the purchase of the whole undertaking.

WARWICKSHIRE ROADS AND HEAVY TRAFFIC.
In his annual report on road maintenance in Warwickshire, Mr. D.H. Brown, the county surveyor, says that it is essential to adopt a stronger form of construction for the less important roads. This step is necessary in consequence of the increasing use of such roads by heavy motor vehicles, including regular omnibus services, which are now running over practically every main road in the county. The provision of wider carriageways is becoming an urgent matter, and the necessary land should be acquired well in advance.

HOW MOST ACCIDENTS ARE CAUSED.
Statistics compiled by the National Safety First Association show that the pedestrian is the cause of most road accidents. The principal faults of pedestrians are set out as follow:- Crossing road carelessly or confusedly 37 per cent., stepping off footway without looking 22 per cent., crossing from behind vehicles 11 per cent., boarding or alighting from moving vehicles 8 per cent., physically infirm 7 per cent., crossing in front of vehicles 6 per cent., failing to use footpath 3 per cent., intoxicated 3 per cent., carelessly boarding or alighting from stationary vehicles 3 per cent.

Offside front door and step, which is lowered by opening the door.

The Seating - A screen at the rear prevents back draught.

[Many vehicles were on offer in the Autumn of 1929, space permits us only to show one]

DAIMLER C.F.6 35-100 H.P. SIX-CYLINDER OMNIBUS CHASSIS

The Daimler Co., Ltd. The C.F.6 has a six-cylinder (97 mm by 130 mm) silent double-sleeve-valve engine, rated at 35 H.P., maximum 100 H.P. having a large diameter crankshaft fitted with a vibration damper. Pressure lubrication of all main, big-end, gudgeon pin and eccentric shaft bearings is provided; also automatic oil-priming and oil-cooling. Water circulation is by pump. Another feature is the coil ignition with spare coil and quick change-over plug. The carburettor, oil filler, etc., are easily accessible on lifting the half-bonnet. A single plate clutch, Ferodo lined and arranged for easy adjustment, transmits the torque via the front cardan shaft with its flexible fabric couplings, to the four-speed gear box. The latter has a casing of a high tensile aluminium alloy, and the sliding gears are case hardened and ground. There is a single lever selector connected to the gearbox through one operating rod. Both halves of the rear cardan shaft are balanced, with Spicer joints at both ends. The front axle is a high tensile alloy steel stamping, of H section between the spring saddles, merging into elliptical section outwards. The swivel axles are of 60-ton steel. Dewandre power-assisted foot brakes act on all four wheels, and the hand brake on the rear wheels. The brake drums are of high tensile aluminium alloy with cast iron liners, the shoes are also of the aluminium alloy lined with Ferodo. The steering is a frictionless cam and roller gear type, the wearing parts being case-hardened. Long front and rear springs are provided, with "Silentbloc" shackle bearings. The body - a double-deck omnibus - is by the Brush Electrical Engineering Co. It seats 52 passengers - 24 in the lower and 28 in the upper saloon - and has open type rear platform and stairs.

Lower Saloon - Brush Double-Deck Omnibus.

Interior of Brush Saloon Omnibus.

MIDSCENE - ROADSIDE CAFES

There are a number of uses for old coaches and buses, many of them ignominious. In many cases their second life is only a postponement to their final journey, although from time to time a vehicle enters preservation after a third or fourth life as, say, a hen-house or summer bungalow.

The Teapot Cafe was found on the A46 but what its internal appointments were like could not be ascertained!

The Bedford ex-Whieldon's was open but the inmates took a very dim view of the photographer's activities and chased him away with a stream of unpleasant language.

The Duple bodied Bedford, too, was open in a layby in Shropshire and after quite a different reception an excellent light meal (bacon sarnies, at a guess!) was had.

Midland Trucking's Bristol was to be found on weekdays parked in a layby on the A422 between Alcester and Stratford during the period when a by-pass around Alcester was being built. Extremely clean and serving good food, presumably the proprietor catered mainly for the workmen.

EVERYONE A LITTLE GEM
A Sampling of the secondhand coach and bus market
1931 - 1991

An often overlooked aspect of public service vehicle operation is that the machines have to be bought. Easy enough for the larger concerns but they 'cascade' their vehicles and always on the lookout for a well-built, well-maintained and reasonably priced bus or coach is one of the smaller operators. In today's terms not many owners of a fleet of two or three vehicles can buy a £120,000 vehicle off the shelf. Even if they could it is doubtful whether their work would justify such a purchase or even that their maintenance arrangements (probably contracted out) could cope with the complexity of their fittings. It must also be admitted that quite a number of such companies' Directors are wary of the problems all too often found on such machines. To quote two examples - early models of a certain round nosed body proved prone to shedding their windscreen pantographs when operating in rough conditions on the motorways. A modification cured this, but although a big firm's finances could withstand such coaches being laid idle, H.P. or lease payments would cripple the smaller man. Another smaller vehicle is, today, superb in its 6-cylinder form, but owners of early 4-cylinder and rather unpowered models found the life of the engines was in terms of weeks rather than months or years. When these were dumped on the market at a give away price, introducing a warning light on the tachograph which operated at 50 mph (80 kph) brought engine life up to a respectable 60-80,000 miles. This was a case where the small man able to run his tours pottering happily on A and B class roads had the advantage over the long-haul operators. The following 'small-ads' are all genuine and are culled from a variety of sources including **Commercial Motor, Coaching Journal, Coachmart, Motor Transport and Coach and Bus Week** and dealers' price lists - in other words, exactly where the small fleet operator would look, then and now. Can we interest anyone in a nice early Leopard? Body is a bit ripe, the engine smokes (only when very hot) but the tyres look good. A real gem...

Seriously though, the vehicles are chosen to give the 'flavour' of the time; and to show the variety available.

1931/32 The last of the World War I chassis are nearing their end, and many smaller makers are either dead or dying.

A.J.S. 26-seater Saloon, coachwork by Lewis and Crabtree, upholstered moquette, small mileage, perfect every way, this coach and also a 26-seater Lioness must be sold, and low prices will be taken.

BEAN. 1930 Bean All-weather, Scotland Yard type, 16-17 seater coach, fitted 4 wheel brakes and numerous extras, chassis and body in perfect condition, £410; any reasonable trial arranged.

DE DION. De Dion, about 1925, 20 seater charabanc, disc wheels and spare, good pneumatic tyres all round, twins on rear; accept £45 or near offer.

1925 OVERLAND Saloon 'Bus, pneumatic tyres on all wheels 2 spares, dynamo lighting and starting in excellent condition throughout.

STUDEBAKER. £14/10, or near offer. 10-seater Studebaker charabanc, first class running order, nearly new tyres, lighting and starting; must sell.

MORRIS 14-seater Carrier's 'Bus, detachable seats, roof rail and ladder, good condition: £50.

THORNEYCROFT Type A. 1 20-seater single-deck 'Bus, 1926, recently equipped with pneumatic tyres.

DIAMOND T America's most reputed chassis, 6-cyl. engine, Lockhead hydraulic brakes, 4 speeds, 4 wheel brakes, 32 x 6 tyres, fitted new luxury 20-seater body by Newns, with sun saloon, the cheapest new job ever offered: £625 complete.

MAUDSLAY, a number for disposal, year 1926, type ML3, forward drive, with 32 seater saloon 'bus bodies, pneumatic tyres, just off service, well maintained; price from £125 each.

1928 KARRIER 6-wheeler 30-seater Service Saloon Coach, upholstered in red leather, comfortable seats facing forward, ample interior lighting, pneumatic tyres, 10% worn, interior and exterior practically unscratched, fitted dynamo lighting, excellent appearance, exceptional value; hire purchase terms; £225.

SPICERS MOTORS LTD - Special Offers!! 1931 Chevrolet Bedford, Dennis, Leyland Cub and Tiger Commer Halley, Maudslay, A.E.C. Regal, latest type luxury coaches, single and double deck buses for immediate and early delivery; our clients will be interested to know that Mr. E.A. Scott, formerly works manager for Duple Bodies, and for 6½ years coachworks manager for Willowbrook, and recently with Strachan of Acton, has now joined us as designer and coachworks manager; his expert knowledge of the highest class of 'bus and coach construction, on which we are now concentrating, will be at your disposal; send us your requirements; Write, call or 'phone; distance no object.

1927 MAUDSLAY all-weather coach, body by Strachan, painted green and cream, snip, £245. 1930 Model G.M.C. T.19-seater sun saloon coach, mileage only 20,000, condition as new, £345. 1930 Model CHEVROLET six-wheel 20-seater omnibus £135.

AEC fleet of K-type single-deck saloon omnibuses for disposal, ex L.G.O.C. services, maintained to their usual high standard, early new 36 by 6 pneumatic tyres... large stock of spares. £75 each.

MORRIS 14-seater saloon bus, body by Buckingham, guaranteed. 14 seater all-weather coach, practically new, used only two months, taken for debt, cost £600, no reasonable offer refused...

A Fleckney, Leicestershire, operator offered: G.M.C. 1928, T40 model, Buick engine, 26-seater saloon, in splendid condition throughout, also 14-seater Chev., M.O.T. both on public service, available June 5th reason for selling, time-table sold.

The Talbot Garage, Kidderminster, offered "any trial" on a Mulliner fabric bodied 20 seat GARNER at £275.

By contrast the Warwick Motor Engineering Co., recommended £1,050 DAIMER C.F.6, forward-drive 32-seater bus, demonstration model, mileage under 1,000, body by Willowbrook and in grey ready for painting to choice, a really magnificent vehicle, reduced from £1,700.

Others included 1929 Delahaye (£225), 1929 A.D.C. (£225), 1929 Crossley (£225), 1928 Minerva (£125), Gilford (from 1928 at £65 to 1930 at £350), 1928 Leyland Lioness and 1926 Karrier CL both with Buckingham bodies at £250 and £95 respectively. One issue alone of Commerical Motor carried adverts offering over 1,300 vehicles.

Reo/Taylor 1928.

1937

By 1937 most of the second-hand vehicles offered in 1931/2 had gone for scrap, saloons had replaced soft-tops, most Continental and American makes had faded from the scene and the buses and coaches then offered new were themselves rattling to their resting places in the nettle-patch. An oddity to be found in 1937 was the bus-coach, hard economics demanded that in the poorer parts of the country the ubiquitous Bedford ran as a bus in the week and a coach at weekends. Diesels were available, but had had little impact on the small-fleet man, who could easily fettle an early petrol engine but was reluctant to tackle the rather awesome (and unreliable) 'oilers'.

BEDFORD brand-new 1937 model chassis with streamline 26-seater sun saloon body, outswept panels, sliding front door, rear emergency door, large moquette seats, suitable for bus or coach work, certificate of fitness.

1933 BEDFORD 20-seater luxury sun-saloon coach, Thurgood body, Easiway roof, parcel racks, high-back seats in moquette, Dunlopillo cushions, curtains, and all the latest refinements at present painted blue and cream, small mileage, with long stage certificate.

AEC Regal luxury bus-coach, 36-seater, sliding door in front, new sun saloon, sliding roof by Sun Saloons of Birmingham, latest-type parcel racks, redecorated and all panels repolished as new, fitted with low pressure brand-new India tyres.

MORRIS-COMMERCIAL 1933, 20 seater bus coach, redecorated to new standard, fitted with brand-new tyres, sun saloon, sliding roof, fitted with beautiful body by Duple.

MAUDSLAY forward-control double-deckers, four and six-wheelers, full enclosed 48 and 60-seaters, splendid condition throughout, certificate of fitness.
20-seater Star saloon bus, 1928, good condition, certificate of fitness to December, 1939, price £50 or near offer.

LEYLAND Tigers, two 24-seater modern luxury coaches, 1932, bodies by Weymanns, sun saloon, moquette, built-in luggage container and indicators, central heating, rear sliding entrance, half-drop windows, toilet compartment with wash basin, very well maintained machines in first-class mechanical condition.

1936 Leyland Tiger TS7-type 32 seater full observation super-luxury coach, with beautiful body, ramped floor, all the latest fittings, including wireless, run 12,000-odd miles only on coaching trips; immediate delivery.

1934 Dennis Ace 20-seater coach, Dennis coachwork, with sliding door, winding head, moquette seats, latest fittings, including wireless, small mileage, finished in the colours ivory white and two shades of blue; for immediate delivery.

TILLING Express 32-seater service saloon, B10A-type, moquette upholstered seats, good body, excellent chassis, long certificate.

LEYLANDS (P.L.S.C.-type) 32-seater service buses, £25 each to clear.

ALBIONS 1929-30-31, six-cylinder buses and chassis, £25 each to clear.

CF6 DAIMLERS 1929, two for disposal, 26-seater all-weather bodies by Park Royal, just been overhauled and ready for season's hard work, tyres good only used on excursion work, seasonal, moquette high back seats, both in very excellent condition, must sell for new coaches coming in, £160 each or offers.

GILFORD interior drive, 26-seater coaches, ex large operator, good mechanical condition, Duple bodies, moquette seats, from £75.

G.M.C. 20-seater luxury coach, with attractive dome back, sun saloon body, floral moquette seats, and continuous parcel racks, chrome radiator and fittings, almost new 32 by 6 tyres, engine and chassis overhauled, painted and certified, £150.

DAIMLER CF6-type 26-seater saloon, in extra good condition, moquette seats, luggage container, parcel racks, mirrors, curtains, etc. well maintained mechanically perfect, good tyres, £150.

Display MORRIS Dictator 32-seater saloon bus, 1932 model in exceptionally good condition, mechanically sound, body sound, upholstery good, unusual bargain, £250.

1946

represented a totally different era for the would-be vehicle seller and purchaser, quite unlike any before or since. In a nutshell, the demand for anything with wheels far outstripped supply and with only the tiniest trickle of modern vehicles coming through - for it was the age of "export or die" - buses and coaches laid up in 1939 suddenly represented a gold mine to their owners. One advertisement, offering a 1943 Guy chassis stated bluntly that it was "only offered due to exceptional conditions". They wanted £500. A 1935 4-cylinder engine extracted from a Maudslay was on its own worth £150, rather more than the price of the whole coach in 1938.

1932 MAUDSLAY 32-seater coach, seating in moquette, three-quarter winding roof, good chassis, body requires repair . . . £325.

1936 MAUDSLAY 32-seater coach, moquette seats, mechanically perfect, not forward control, good tyres . . . £1,200 o.n.o.

1932 DENNIS Lance 6-cylinder 32-seater buses, bodies need repair, no seats, bargain price £150 each.

1934 THORNYCROFT 32-seater coach, fitted with Thurgood body, red leather interior, engine and chassis 100%, but needs re-painting £850 o.n.o.

1938 Leyland Cub SKZP2 [1935-38] 26 seater, Burlingham body, sun roof to rear, all in excellent condition £1,800.

Motor Coach 26 h.p. CHEVROLET 20 seats (removable) perfect runner, suitable canteen etc. £350.

DODGE 20-seater saloon, unlicensed since 1938 and little used.

LEYLAND LT5A [1933-34] 36-seater rear entrance service-bus bodies only; these are Leyland-built bodies, tubular seating, from £200 each.

Only another dozen vehicles were individually advertised but Murphy Brothers of Willow Street, Leicester had around ten for sale in July ranging from 1932 THORNEYCROFT, AEC Regal, BRISTOL and ALBION to a 1939 LEYLAND Tiger, but all were priceless . . . while Don Everall, Wolverhampton had, incredibly, a 1930 Crossley 32-seater with "recently overhauled engine" at £350.

Austin/Barnaby 1948

1955 and the golden days of vehicles sales are gone and many say with them went the golden days of coaching, although there were another five years during which a good operator would continue to find demand for seats outstripped his vehicles capacity. But by 1955 the early post war coaches and buses, all too often with by now 'ripe' bodies were becoming available, and there were bargains for a number of impecunious, stupid or unlucky men had overstretched themselves with their purchases and even a day or two's lateness in paying installments could lead to a gang of heavies, quite legally, removing a coach from the yard (or garden, more often!); it then being sold at a "knockdown" price sufficient to meet outstanding debts. The similarity of these H.P. firms' activities to those of today's banks and building societies is quite marked. "Used Passenger Vehicle" small-ads totalled about 400 per issue of "Commercial Motor" and another 200 vehicles often more upmarket were to be found in "Coaching Journal". Virtually all chassis/bodywork combinations were available; those below give an impression of value-for-money, particularly when read against 1946.

1950 AEC, 33 seater Burlingham, selector 9.6 low mileage, immaculate condition, only £1,500.

Two April, 1950, special BEDFORD 18-seat luxury vehicles, low mileage, excellent condition, £1,025 each.

1947 BEDFORD Duple sun saloon 27-seater Dunlopillo, 3 months guarantee. H.P. terms. £545.

£1,475 only. 1951 (August) BEDFORD Duple Vega 33-seater luxury coach, courier's seat, radio and heater, one owner since new, been carefully used and maintained. Box No. . . .

1955 COMMER Duple 41-seater coach, fitted TS3 engine, mileage 30,000, available after Jan 1st 1956, £3050.

CROSSLEY Luxury coach, 33 seater, good looker . . offers. Reason for selling giving up for commercial transport. [And another, similar] 1949 CROSSLEY . . . disposal due lack of space £1100.

1945 DAIMLER low-bridge double-decker, reconditioned 7.7 engine £475.

1948 FODEN . . . luxury coach by Associated Coachworks.

1951 LEYLAND . . . Gurney Nutting coach body

1950 TILLING-STEVENS . . . full-front Bellhouse Hartwell coachwork

1951 LEYLAND ROYAL TIGER . . . 39 seats observation coachwork by Whitson

1948 CROSSLEY mounted with Dutfield coachwork . . .

1953 BEDFORD with coachwork by Gurney Nutting . . .

1950 COMMER Avenger mounted with 33 seater coachwork by All Weather

1950 GUY Vixen with full front coachwork by Devon Coachbuilders

1947 DENNIS J3 fitted with Santus body

1952 FODEN Metalcraft 41-seater body £2,660

1951 DENNIS Falcon, 33 seater Reading body very clean £1,600

1950 MAUDSLAY 33 seater Gurney Nutting luxury body £1,875

1949 AUSTIN Pearson, 28 high-back seats £550

1950 AUSTIN, 32 seater Kennex full front £875

1949 AUSTIN Mann Egerton full front bodies, most attractive vehicles £825

1946 AEC 7.7 diesel, 33 seater Plaxton body £495

1948 COMMER Commando, Plaxton 30 seater, suitable works service £195

1954 SENTINEL/Duple Elizabethan £3,000

And so the offers went on, Crossley/Streets, Leyland PSI/Bellhouse Hartwell, Bedford utilities (from £400), Commer/Churchill, Crossley/Strachan; Maudslays galore, Crossley/Trans-United (snip £650), a rather sad offering from Vincent Greenhous, Shrewsbury of a Bedford **Opel** Duple bus, without seats, suitable for travelling shop, Don Everall had 40 vehicles in stock, P.V.D. Marton near Rugby, 80; Lees Motors, Worksop only 7 but including a Borgward micro-coach "with Plaxton interior".

1972 and the secondhand market reflects the 'Britishness' of the contemporary coaching industry, albeit both Ford and Bedford are, as we know but like to ignore, American owned companies. A surprising number of Midland firms are advertising used wares including Erringtons of Evington; Don Everall of Wolverhampton; Yeates, Loughborough; Coach Sales Consultants, Loughborough, Moseley of Shepshed and Gloucester and Vincent Greenhous, Hereford.

Coaches have a greater capacity and more refinement in the way of heaters and p.a. equipment is offered. That said, the old, almost Victorian, ambience of polished wood, antimacassars and carpet is replaced by plastics and big windows, but Telma electric brakes help the driver.

coach, heaters, radio and public address, microphone, £4,250.

BEDFORD 1963 SB5 Duple Bella Vega 42-seater, radio and public address, black and cream £2,150.

1968 BEDFORD 52-seater VAL70. Plaxton from £4,350.

1968 BEDFORD 52-seater VAL70 Duple with Telma £4,400.

1964 BEDFORD SB5 Duple Harrington Crusader 41-seater, radio and p.a. £2,400

1964 BEDFORD Val 14 Harrington Legionaire 51-seater, radio, p.a. £2,500

1956 FORD 52-seater Duple £3,100.

1965 FORD Duple 42-seats, radio, public address, seats re-trimmed two years, short engine fitted November £2,400.

1964 FORD Plaxton 49-seater, forced air ventilation, radio, wheel disc, recently repainted £3,600.

1965 FORD Duple Firefly, 41 seats, first class condition, red interior, 5-speed gearbox, two heaters, mechanically very good, offer, £2,200.

1971 BEDFORD Dormobile 11-seater PSV black ambla seating £1,050.

1970 FORD TRANSIT Dormobile 12 seater PSV black and ambla seating £900.

1970 COMMER 13 seater PSV red leatherette seating £895.

1968 B.M.C. 12 seater PSV brown leatherette seating £650.

1955 BRISTOL, Lodekka, Gardner or Bristol engines, with or without platform doors, from £475.

1959 LEYLAND Atlanteans M.C.W., 78 seaters, high bridge; also 73 seater, low bridge, £1,400.

1957-59 BRISTOL SC 35-seater service buses, Gardner 4LK engines from £450.

1962 ALBION Nimbus, 31-seater Harrington, O.M.O. £775

1963 A.E.C. Reliance (front entrance) Plaxton Panorama 41 seater, interior red moquette; extras include 5 speed gearbox, heaters, etc. £2,550.

1982 We have now passed into the age of vehicles recognisable today, for many of these offered here, particularly those being disposed of by Shearings, Cotters and the National Group, and so on are those which today lead an unhappy afterlife as 'Contract' or 'Schools' vehicles. Few vehicles are for sale with bodies other than Plaxton, or Duple, unless they are continental confections, for by 1982 it was almost 'de rigeur' to buy foreign, operators perceiving that to buy British was somehow lowering.

1977 FORD R114 Plaxton 53 seater £14,000.

1978 FORD R114 Plaxton 53 seater £16,000.

All these coaches are in excellent condition, with P.A. systems, arms to seats, illuminated side panels and additional trafficators.

1971 AEC Swift Marshall 53 seats, single front door, £3,000.

1977 BEDFORD YMT 53 seater, radio, P/A, Telma, short motor fitted August 1980, £15,000.

1979 BRISTOL LHS 33 seat Plaxton, radio P/A, blinds to side windows, exhaust brake £17,500.

1973 BEDFORD YRT Caetana 53 seats £4,150.

1973 FORD R1014 Duple Dominant, 53 seats, £3,500.

1972/3/4 LEYLAND PSU3B/4R Telma. Plaxton 49/53 seats, armrests, pleasing interior, automatic lubrication, Kysor blinds, according to age £6,750 to £11,250.

1980 VOLVO B58, 12m Moseley Alpha 53-seater £40,000.

1980 FORD A.0609 Faro III 25 seater £17,000.

1980 FORD Transit Moseley 'Envoy' 12 seater £8,000.

1975 BEDFORD YRT Moseley Estoril 11 53 seater blue/pink £13,000.

1975 BEDFORD YRT Plaxton Panorama Elite 53 seater cream/red £13,500.

1973 MERCEDES Diesel 23 seater coach £3,250.

1973 SEDDON Pennine VI Plaxton Elite 49-seats £6,500

1973 B.M.C. JU250 12 seater, blue £550.

1970 BEDFORD VAM70 Duple Viceroy 5-seater £950.

1966 BEDFORD VAM 14 Duple Viscount 45 seater £700.

1992 Most of the hallowed British names are now coming to an end, Bedford, Ford, AEC, Seddon, MCW, are all gone as chassis manufacturers and in another decade Leyland will only be associated with tired works, contract or school buses. Almost alone are Dennis in fighting back against the Continental invasion, albeit incorporating (in the main) American Cummins engines. Although Mercedes and other imported or designed chassis are used for the basics a surprising number of British bodybuilding firms cater for the 12-30 seat market, while the big boys are all building "Green" people-kindly bus bodies. Sales of new vehicles are, however, almost unbelievably low, so that good second-hand coaches and buses still fetch a reasonable price. Tatty ones don't! But despite the depression and uncertainty found in the early part of 1992 it proves something that at one of the West Midlands Traffic Commissioners seminars for new operators 30+ representatives turned up and most would be in the market for "a little gem".

1990 H Reg MERCEDES 408, 15 seat minibus £17,250.

1990 H Reg MERCEDES 709, 21 seats, 49,750 kms only, £22,500.

1987 FREIGHT ROVER Sherpa long wheelbase, 2.5 diesel, 20+2 standing, electric entrance door, £3,750.

1988 TOYOTA Optimo 21 seats, radio PA, red interior, immaculate condition, £22,500.

1990 FORD transit 130 diesel, 14 seat Deansgate conversion, radio cassette, soft trim, saloon heater, roof vent, £10,750.

1984 BEDFORD YNT 500 Turbo, 6 speed ZF Duple Laser (53) power door, radio & PA, speed limiter £19,000.

1986 LEYLAND Tiger Duple 340, 49 seater plus toilet, Gardner 6 LXCT, £29,850.

1981 LEYLAND Leopard, Duple Dominant 11, 53 seater, £10,500.

1983 VOLVO B10M Jonchkeere P599, 49 recliners and courier seat, sunken nearside toilet - a clean machine, ready to drive away. Best offer will secure.

3 LEYLAND Atlanteans P.Reg. double deckers, red moquette seats £3,750 + VAT each.

1960 AEC Harrington Cavalier ex Neath and Cardiff Coach . . . £2,500 + VAT.

1975/6 VOLVO Ailsa deckers, tidy but requires slight mechanical attention, drive away £2,495.

BARGAIN BASEMENT . . . 1980 VOLVO B58 Unicar, 11m, 52 seats, 6 sp ZF, tidy interior, body requires attention £6,950.

1986 NEOPLAN Skyliner (Mercedes/ZF) 71 seats, toilet, servery . . . offers around £39,500.

1982 X BRISTOL LHS Plaxton Supreme V 31 Lazerini recliners, tinted windows, curtains, power door, recently repanelled, excellent condition, £12,500.

LEYLAND Lolander, 1965, 71 seater, £1,950 o.n.o.

This last really must be a "little gem" indeed, albeit more of interest to a preservation group than a hard-nosed operator. Throughout this survey the advertisements are real, although I have shortened them here and there. One day, if I am permitted, I will tabulate who had what for sale when; results could well be interesting.

Leyland/Park Royal 1976

Bristol/NCME 1969

An often necessary adjunct.

Dennis/Gurney Nutting 1952

Leyland/Leyland 1929

Leyland/Bellhouse Hartwell 1952

Thornycroft/Clark c. 1930

AEC/Duple 1958

Leyland/Duple 1951

AEC/Yeates 1960

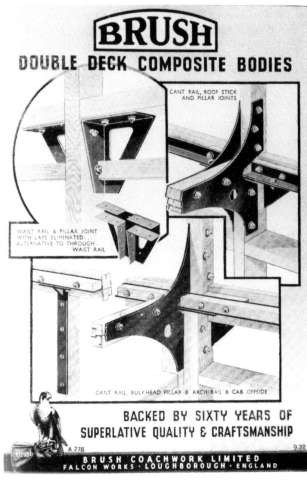

GOLDEN DAYS OF ADVERTISING

These nine advertisements cover roughly the period 1935-1955. The Daimler text gives a feeling of deja vu as they were glad then in 1935 to sell 6 vehicles - in the early 1950s this was worth little, orders for fifty being quite normal - but yet again, Coach and Bus Week gets quite excited in 1992 when an order for 6 Volvo or Dennis vehicles is placed. The semi-utility bodywork shown in the Guy advertisement dates this as surely as the '1946' of Brush's offering. Interestingly the Maudslay advertisement was used during the war with different text. Then they carried workers to keep factories open, here export was the keynote.

82

for ease of maintenance

—and routine inspection, the underfloor diesel engine of the Sentinel 30′ × 8′ passenger chassis would be hard to beat. It is a six-cylinder, direct injection unit and has a five speed gearbox to ensure smooth, powerful traction whatever the road conditions. For short service runs or long distance luxury travel the economy and reliability of the Sentinel 30′ × 8′ is unrivalled.

ENGINE SPECIFICATION
Sentinel six-cylinder direct injection diesel engine.
Capacity . . . . . 9.12 litres
Bore . . . . . 120.65 mm.
Stroke . . . . . 133.3 mm.
Compression ratio . . 16.75 : 1
Max. torque at sea level
 375 lbs. ft. at 1,000 r.p.m.
Max. b.h.p. . 120 B.H.P. at 1,800 r.p.m.

and passenger comfort

The beautifully sprung 30′ × 8′ allows comfortable seating capacity for up to 44 passengers. Advanced design of suspension and springing gives freedom from vibration at all times, and the underfloor position of the smooth running engine ensures that no noise, heat or fumes enter the cab or body.

Sentinel

SINGLE DECKER 30′ x 8′ PASSENGER CHASSIS

SENTINEL (SHREWSBURY) LTD. SHREWSBURY, TELEPHONE: SHREWSBURY 2011

THUMBNAIL SKETCH 4 - BURTON-on-TRENT CORPORATION

Initially served by an orthodox town tramway Burton-on-Trent also boasted an interurban, the Burton & Ashby Light Railway. The former closed in 1929 and the latter in 1927.

From a transport manager's angle Burton's great charm as a bus town lies in that, to all intents and purposes, it is level. The problems were that until recently the roads were cobbled and criss-crossed by railway lines feeding the breweries that were Burton's lifeblood. Timekeeping was always a bad joke and drivers became adept at finding ways round brewers' drays, loading, unloading or just stuck.

The Corporation obtained powers to operate motor-buses in 1916, but did not commence services until April 1924. Then, and until the company could no longer provide the short (27'/8.23m) chassis preferred by the Corporation, Guy Motors Ltd chassis were used almost exclusively; totalling around 145 between 1924 and 1962. Quite remarkable longevity could be attained with the Corporation workshops at Horninglow Street thinking nothing of totally rebuilding bodywork, typically ex-London Transport Guys with Park Royal & Weymann bodies supplied to LT in 1945, mildly overhauled by Roe of Leeds in 1953, and fettled by the Workshops ten years later were still serviceable until 1967 - one is preserved. The first metal framed 'Orion' bodies arrived in 1957, 8'0" (2.44m) wide vehicles in 1961, and although they perforce changed to Daimler chassis for 1962 deliveries, the relatively rare CSG5 (David Brown built synchromesh gearbox) model was chosen, with thereafter the even more improbable CCG5 (constant-mesh Guy gearbox). Fleetlines first arrived with a single deck demonstrator, then the first with North Counties double-deck bodies came in 1970, followed by a further batch (albeit with 'Nottingham-style') Willowbrook bodies in 1923. Not many single deckers were operated postwar, but they did include a couple of Burlingham bodied buses ex Bournemouth. On 1 April 1974 the operating authority changed to the East Staffordshire District Council; the loss to Burton of its independent fleet was immeasurable.

FA 8033 Guy Arab II 5 LW/Park Royal in service 1945 - 1964

FA 9714 Arab III 5LW/D.J. Davies in service 1950 - 1969

FFA 471 Arab IV 5LW/MCW in service 1957 - 1973

NFA 876 Arab IV 5LW/Massey in service 1961 - 1973

RFA 406J CRG6LX/NCME in service 1970 to ESDC

NFA 19M CRG6LX/Willowbrook in service 1973 to ESDC

SFA 83 Daimler CCG5/Massey in service 1963 - 1973

TRAINING DRIVERS
The Activities of a Northampton Concern

Bus operation, regular long-distance coach operation and private hire work are covered by the firm of Midland Motorways (Nightingale and Sons), of Northampton, which is managed by Mr. F. Nightingale, senior partner and a son of the founder who started as a horse job master and still takes an active interest in the business. At present the fleet, which includes Guys, Gilford, Maudslays, Studebakers, Brockways and a Reo, consists of eighteen vehicles.

Approximately 1,000 miles per week, winter and summer, are averaged on private hire and excursion work alone. Football matches and races are catered for during the winter, and in the summer individual bookings by middle-class people, including many private motorists, for advertised day trips and extended tours, contribute substantially to the revenue. Throughout the organisation the personal touch is noticeable. Mr. Nightingale attends to the bookings for these trips, superintends wherever possible the loading of coaches, and when a job is large enough to necessitate a convoy of coaches, himself accompanies the party. He has found this is much appreciated by his clientele. The rougher element is not encouraged but even so a certain amount of difficulty is experienced with cheery passengers who insist on trying to persuade drivers to drink, but drivers are definietely forbidden to drink during working hours. In general it has been found that rowdyism is dying out. Another difficulty, but of a different kind, has been that experienced with village choir and similar outings, who often want a 24-hour day from 4 a.m. to 4 a.m.

It is Mr. Nightingale's opinion that many troubles of the bus business are due to lack of care in selection of drivers, and he accordingly goes to great pains to ensure the efficiency of his own. A new driver has to pass a private driving test as well as that imposed by the local police, and is then asked questions to see that he is capable of dealing with any running repairs that may prove necessary on the road. If he survives this, he passes a short period on the local bus services under personal supervision. From the first the slogan "as soon as you stop look over the vehicle" is drilled into him, so that many embryo road stops are nipped in the bud. Attempts are also made to keep every driver to his own coach, although, of course, this is often impossible. The better coaches are always reserved for long journeys, and their drivers are held responsible.

The question of conductors is also engaging the attention of the firm. At present women conductors are employed on the local bus services and have proved altogether satisfactory. During the rush periods male conductors are employed for the London-Northampton coach service, usually one conductor to two coaches running together. This is to relieve the driver of ticket-issuing duties when passengers without tickets board the coach en route. Further to save time in view of the 30 m.p.h. limit, Mr. Nightingale has devised a railway-type ticket to be issued by the driver, but not by the agents.

MOTOR TRANSPORT 12 January 1931

TOP TO BOTTOM ... AND BOTTOM TO TOP

In general it was always a belief in the omnibus industry that the ideal employee was one who started at the bottom and very often reading Annual Reports of Midland companies and corporations one comes across examples of the boy who on leaving school became a 'Parcels' Lad', carrying out deliveries either on foot or bicycle. On a tramway he might start as a 'Points' Boy', or as the 'Young 'Un' in the workshops. Not even, perhaps, as an apprentice but a 'rubbing rag', carrying out odd jobs. Progression was then to conductor and later driver. If the boy attended diligently at his night-school and if he made himself known to those who, effectively, owned him, if he did not drink, if he attended church regularly and if he was lucky, he could then progress up the ladder. But how many did succeed out of those who started on the bottom rung? Marriage could cut two ways; a married man was considered 'steadier' but the hours and conditions of work destroyed (or at best weakened) many a set of vows.

Even as long ago as the mid-1930s a traffic manager stated "The qualities and knowledge which might be expected from the ideal transport employee would be as follows: (1) honesty, courtesy and personal pride; (2) a knowledge of arithmetic and writing, and general mental alertness; (3) salesmanship; (4) a knowledge of the principles of engineering; (5) a knowledge of the law as it affected transport; (6) an elementary knowledge of economics".

Quite remarkably, in an age when every employer believed himself to be capable of selecting the best, a heretical suggestion was put forward by another Traffic Manager. "In order to start off with the right material, the applicant for a position [of conductor/guard] should be interviewed by a psychologist, who would ensure that he would have the requisite qualities". He added that further tests would have to be passed before promotion to driver, to show "his mental alertness and quickness in avoiding accidents". He also recommended retirement at the age of 60. All quite remarkable when one considers this was written in the mid-1930s and only now are psychological tests being adopted with any degree of enthusiasm; a new driver at one local company has a 2½ hour written test and one hour on a computor answering random questions. The pass mark is, one gathers, adjusted to meet the ratio of applicants to staff shortages. But drivers still quickly leave as a combination of shift hours (particularly as split shifts are on the ascendant) and the public attitude takes effect.

Whether a man (or woman) could get a commendation depended both on his employer and the location of his depot. A driver for a small Northern firm in the 1950s and 1960s commonly worked a 70-hour week, in fairly awful conditions (it's no joke to sit under a leaking roof and to have snow blowing around the feet for 8 or 9 hours) and expect (and get) no thanks although an incentive pay rise of 50p a week might be forthcoming after a year or two. On the other hand Corporation or City transport employees were more directly appreciated. In Transport World, 11 November 1943, appeared a citation, partly designed to encourage others, but genuine enough: ". . . the majority of transport employees are pulling their weight. A shining example of loyal service of this character comes from Coventry. A Corporation bus driver, so I am told, rises every morning at 3.30 a.m. He cycles 12 miles into Coventry in the morning to get to his depot, and pedals back the same distance every night. Only twice this year has he been late, on both occasions this was due to his bicycle breaking down. Mr. R.A. Fearnley, general manager of the Coventry undertaking, has rightly congratulated him on this fine example of devotion to duty". One problem today is that the older men who, rightly or wrongly, gave and expected loyalty from their employers are being replaced by the 'quick buck' brigade. But too many employers ruthlessly hire and discard their men at the instruction of their accountant with little or no interest in what happens to the staff when they are paid off. Back in 1964 H.F. Warner, of Warner Motors, Tewkesbury (now Warners Fairfax Tours) described how he, as one of the pioneering coachmen, saw his company: "We're all-in men here. The driver who's decorating the office next door returned from a Paris tour last week. Coach drivers on tours like something useful to do in between times and off-season. One of my sons is the engineer and the other is the traffic manager of the company, but they both drive school and works buses in an emergency. The mechanics drive as required. So do I on occasions and we fit in the shoppers' stage-carriage services without much bother. I've been driving for 50 years - since 1913 that is. Some of our men have been with us for over 30 years".

This appeared in Commercial Motor on 31 July and it is true to say that Mr. Warner might (to a reader in the 1990s) as well have been writing about another world. But to another operator in 1964 this was no more or less than was expected.

During two wars companies were pleased enough to employ women bus, trolleybus and tram drivers, albeit with some reservations about their strength and implied (if not expressed) doubt over their ability to withstand the hours and conditions. In the end, people are all individuals. Some women did well, others poorly. But just over 20 years ago it was reported in Commercial Motor (30 June 1972) that the National Industrial Relations Council could do nothing to help a Midland Red conductress, Mrs. Doreen Barnett, in her wish to become a driver at the company's Oldbury garage.

Mrs. Barnett had applied to the court because after a ballot at the garage it had been voted by members of the Transport and General Workers' Union not to allow her to become a driver. Midland Red already employed 13 women drivers at other garages.

Sir John Donaldson, the court's president, said there was nothing under the law NIRC could do to help but said that Mrs. Barnett had been quite right in applying to the court. Mrs. Barnett's claim was dismissed. Incredible really, this wasn't 1872 but 1972. And, if we look back to the Bus and Coach issue of January 1962, still only just over thirty years ago we find a matter of-fact comment that "Every day bus drivers and conductors can be seen smoking on buses or coaches which are carrying passengers. Employers and employees know this is illegal, yet in some cases no efforts are made to stamp it out.

For some bus operators know that, with the staff situation as it is, to dismiss an employee for an offence such as this merely means that one more bus does not run. So if disciplinary action was taken, both the operators and the Traffic Commissioners would be beseiged with complaints about buses which did not run. The police, too, usually turn a blind eye to drivers smoking.

Any law that either cannot be enforced or is not enforced is a bad one. It may be thought undesirable for a bus driver to smoke while driving, but in present-day circumstances it obviously cannot be regarded as so very reprehensible. After all, many motorists smoke while driving.

What is dangerous is for the driver of a large passenger-carrying vehicle to hold the steering wheel with one hand while trying to strike a match with the other. Surely, then, it would be more sensible to insist that drivers light cigarettes or pipes only when their vehicle is stationary, and to enforce that policy".

To refer back to the beginning of this article. Our traffic manager stated he expected an employee to have the virtues of "honesty, courtesy and personal pride". Did he, do you think, mean this driver who achieved fame in Transport World, 11 November 1943?: "In a recent court case a bus driver successfully appealed against a conviction and fine for failure to behave in a civil and orderly manner towards a passenger, a taxi driver. His counsel said that bus and taxi drivers were "prone to look upon certain words in a different light to other people".

An employer with only a handful of staff to worry about had the disadvantage that if one went sick or upped and left then there were problems (often overcome by his wife having to drive!) but at least most drivers developed some degree of loyalty and the Boss by knowing their foibles could humour them. A big company or corporation transport manager took pot luck.

In March 1948 the craftsmen and maintenance men employed by Coventry Corporation went on strike when they were refused a 2d (0.83p) per hour rise in pay. Moral blackmail was applied as the Corporation announced in the press and through the radio that if the strike continued vehicles would be withdrawn and services cut because of the risks involved in running non-maintained vehicles. The old-fashioned British compromise was arranged!

But less than two years later the General Manager of Coventry Corporation Transport, Mr. R.A. Fearnley, was complimenting two men as being "real stalwarts of the transport undertaking". One was Mr. W.R. Mumford the works superintendent. And his "quite remarkable" record serves as an example of how an 'ordinary' employee could progress. He had completed 50 years service starting "as a boy of 15 in 1899, when the transport strength was four trams and the system was being converted from steam power to electric traction. He rose to the position of being in charge of the maintenance of tramcars, and later motor buses, and since 1942 has been superintendent of the Watery Lane works". The other "stalwart", Mr. F.O. Brand, had, if anything, an even more unusual record. His 42 years had seen him start as a conductor, very quickly become a driver (in 1908) and after front line service in World War 1 he joined the garage night staff, the most soul destroying job of all. "As superintendent of the night garage, he had a busy time during the blitz of the last war, when his work was in the target area. During the whole of the war period" said Mr. Fearnley, "the despatch of buses every morning failed only once - the morning after the November 1940 raids".

Bearing in mind that throughout the war they worked at least 12 hour shifts, were on the fire-watching teams (looking out for incendiary bombs), had the racket of war around them to destroy sleep yet were still loyal to their employers made them, indeed, quite remarkable men.

IT IS IN YOUR OWN INTEREST TO HAVE THE GOODWILL OF THE PUBLIC

THIS CAN BE ATTAINED BY:-
  GIVING HAND SIGNALS.
  STARTING & STOPPING SMOOTHLY.
  APPLYING BRAKES GENTLY.
  PULLING INTO KERBS AT STOPS.
  KEEPING A SHARP LOOKOUT FOR INTENDING PASSENGERS.

MIDSCENE - PEOPLE

"Bessie". 1 November 1915.

1940s, said to be Leicester.

1915-1919 "Derby".

The Aldred family

Horsley Woodhouse pre-1934

The Gem

Unknown, 1920s Derby area

Service 1930-1932

"Eagle Service", Little Eaton c. 1934

The humour of busmen and buswomen had always varied between the downright crude to relatively sophisticated. Jokes from outside the industry vary from the vicious to the sympathetic. A brief selection is given below; the preservationist looks rather familiar somewhere.

****

When women drivers were first introduced into the 'Midland Red' fleet a scurrilous and probably apochryful story went round to the effect that one of these souls, being unable to complete her knitting in the canteen, was sighted by a policeman near Lutterworth busy at work with the fingers while controlling the steering through her wrists. Drawing alongside a police officer shouted to her "Pull over". "No", she replied, "socks"!

****

Overheard at a Walsall bus shelter. Inspector to youth, who was leaning on the outside. "What am yo' doing?" "Nuthink". "Well yo'm stop it yo hear".

****

As a woman was climbing on the bus she turned to wave farewell to her friend. "Good-bye, dear" she called, "I'll be with you again soon". "Sooner than you think, lady", said the conductor, firmly. "We're full up".

****

Q. How many passengers can you get into an empty bus?
A. One. (It then ceases to be empty).

****

You may have heard about the lady coach passenger who lost her false teeth? Apparently the particular coach she travelled in had a peculiar swaying motion and she was sick. Sure enough, she claimed compensation for the loss of one perfectly good dental plate. On solicitor's advice the company paid the claim.

****

NOTE: The times shown on this Time Table are only an indication that the vehicles will not leave the Scheduled points before such times, and the Company do not undertake that the Omnibuses shall start or arrive at the exact time specified.

****

Overheard one Market Day in a Border town.
" 'ow you'm get 'ere so early, ba Jim?"
"Ah. Doris did pick we up. Goodish load, but 'er squoze we in".
Doris was one of our local bus drivers, around 6'0" tall, and weighing over 15 stone. When she physically squashed passengers on they stayed squashed. The last person to board usually stood on the step and acted as the door. Even at 30 m.p.h. this could be interesting.

****

PERILS OF A PRESERVATIONIST

The advertisement claimed "one previous owner" but I didn't know he came with it!

WE STOPPED AT ALL THE PLACES OF INTEREST ON OUR CHAR A BANC TRIP!

TICKETS FROM A CONDUCTOR'S EYE

Pensioners, particularly those travelling on free passes, have led to a distinct language building up among O.P.O. bus drivers. Generally the name 'wrinkly' is applied, but where these passes have a start time (usually off-peak) this title changes to 'twirlies'. The derivation is obvious: "Am I too early [to use my pass?]", a phrase heard by every driver every day. Strange anomalies exist, as when a bus leaves the terminus at, say, 0858, and the passes are not valid but two minutes and one stop later all the pensioners can board. The underlying ideas are to avoid overcrowding in the peak period and to reduce the chance of these pensioners using them for 'gainful employment'. One day an elegant woman got on five minutes before the validity of the pass. The driver, a 5 a.m. man, already had glazed eyes and would have passed a bingo ticket. The pensioner was followed by an inspector who charged her the 70 or so pence fare. Another passenger said this was hard. "This lady, Madam", answered the inspector, "is an employment officer at a large Department store and earns more than you'll ever see". The would-be fare evader blushed.

But if you want to make a pensioner happy, stop her on the platform and closely inspect the pass looking from photograph to holder and back. By now she'll be hopping from paw to paw. Give the pass back and tell her if you catch her using her mum's pass again "Action will Be Taken!" It makes their day.

An anomolous ticket that dragged out its' existance until the 1970s (and may still exist here and there) was the old Workman's Single or Return. All of these varied in their availability from municipality to municipality and company to company but were often inherited from tram days. At its simplest the Workman's ticket was only issued to "Artisans and Persons of the labouring classes" but Company byelaws put the definition rather negatively. For example our company stated that "A passenger, not being an artisan, mechanic or daily labourer within the true intent and meaning of the Acts of Parliament relating to the Corporation, shall not use, or attempt to use, any ticket intended only for such artisans, mechanics, or daily labourers, or travel or attempt to travel on the company's vehicles as an artisan, mechanic, or daily labourer".

Obviously there were gross anomalies in this system for if an artisan were to rise to Foreman, by virtue of wearing a tie he immediately lost this privilege and there were genuine cases where promotion, given this increase in fare and the necessary improvement in the rented house, caused foremen to be worse off than previously. Similarly many of the mill's office girls we carried had to travel at the same time as the machinists, earned much, much, less but paid full fare - as much as a shilling single against a fourpenny workman's return. (Try 50p against 15p). Outward travel times were always restricted, usually prior to 6.30 a.m. or so, but again these early morning vehicles cost the company extra money for drivers and conductors (time before 6.30 being unsocial and earning an extra shilling or two), as well as the vehicles often needing to be cleansed of coal-dust or clay-stains before "them in 't' office" could travel at 8 a.m. One morning I had a typical Black Country conductor. We'd started at 4 a.m. on the Monday and were just having a warming bowl of tea when a diffident looking clerk approached us. I agreed we were on the bus. He explained he'd been to wet the (new-born) baby's head during the weekend and had only enough for a Workman's Tuppenny Ticket but could borrow from his mates at work for later in the week. "Ar", smiled my mate, "let's have a look at yo honds". Puzzled the lad held them out. "Yo bin in gardin ave yo?" The lad agreed. "Ar reckon that meks yo a werker for me" and he gave him his ticket.
It sort of met the rules!

"A conductor ... shall not wilfully deceive or refuse to inform any passenger, or intending passenger as to ... the fare for any journey".

"How much to Smith Street?" "6d". "How much to Jones Street [¼ mile nearer]?" "6d". Exasperated, "How much to Brown Street [½ mile nearer]?" "Still 6d". "Oh, bother it, I'll get the next one" and, presumably so she did, for a day or two later a visit to the office on my part was desired. The Office Person tutted and tished while I tried to explain. At last, feeling quite aggravated I pointed out there was a massive sign on the side of the bus that explained the whole thing. "Route 66, Sir, has a 6d minimum fare, Sir". "Ah!"

But then as we say passengers "Enter bus to make fuss". Did you know railwaymen have a similar saying "Enter train, disengage brain"?

When buses used to be really crowded quite often there was no way a conductor could get through to issue tickets and 1d and 2d coins would be passed over from hand to hand, perhaps the whole length of the bus. One day a florin arrived in my hand, passed by a beer-smelling builder's labourer. A 2d ticket was required. I gave him the ticket and warned him there was 1/10d in change. I watched as that ticket passed through at least 10 pairs of hands before reaching the clerk right up behind the driver. Says a bit for working men's honesty that the change also reached him intact.

Conductor 0129

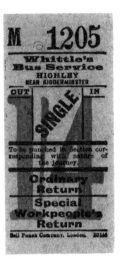

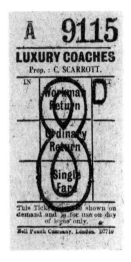

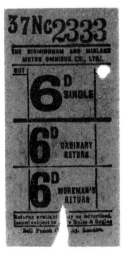

This Ticket may be used for a 2½d Workman's Return Journey. It must be tendered to the Conductor for cancellation during both single journeys on which it is used, and shown on demand to any official of the Department. Both journeys to be taken on the same day by the same person; the first journey before 8.30 a.m., and the second after 4 p.m., weekdays only.

THE NUANCES OF TICKETS

Tickets are a mixture of the mundane and the arcane. Mundane because to 99% of passengers, particularly on buses and only slightly less so on coaches, a ticket is a ticket. The mystery of tickets though lies in the secondary activity. Having given a receipt (the ticket) the issuer then wants to check how much revenue has been gained. Very early on bus companies grew to distrust conductors and various methods were proposed for checking tickets issued against the waybill (which theoretically shows issues) and the cash passed over. The Bell Punch type of machines was the simplest as the 'ting' indicated the cancellation of a ticket and the little coloured discs cut out could be extracted from the hopper and counted. One girl who was employed at this job had, previously, worked in a factory where she counted the number of dried peas in random packets as they came down the line. She preferred counting the discs as it was only one part of her days work and, furthermore, she was sitting down. Williamson ticket machines chopped off the unwanted residue of a pre-printed ticket which could then have its value calculated. Bellgraphic used a carbon to record issues and Setrights add up the total issues in £ + p mechanically. Electronic machines are far simpler, expensive and prone to odd maladies. For example their memories, if not the body, can be (and have been) wrecked by exposure to microwaves. And a tishy bit of paper isn't really a 'proper' ticket is it?

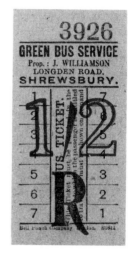

2545 Rose. Reverse: **G. WOODWARD** MALVERN & CRADLEY BUS or CHARA OUTINGS any distance 'Phone - Suckley 18

3926 Gold. Red overprint. Reverse blank. Unusual to show "Omnibus ticket"

121 7147 Dirty pink. Red overprint. Reverse blank
07254 Mid grey. Reverse blank

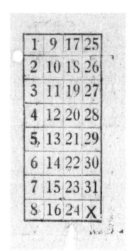

8A 8079 Dull gold. Green 'stripes'. Perforated ticket. Reverse: Always **GOODWIN'S EXTRA SELF RAISING FLOUR**. It must be - The Baby Picture Bag. Advt. No. 0966

14 Bb 2570: Grey. B.M.M.O. 1/- triple ticket single/ordinary return/workman's return. This is reverse. Rather unusual.

8 Dr 5454 Gold. Reverse: **KEATING'S KILLS** Beetles, Moths, Fleas even Bugs Cartols 6d & 1/-

B2390 Dull green. Reverse: JONES' COACHWAYS LTD. MARKET DRAYTON

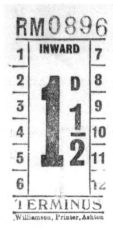

1603. Cream, reverse: The **LUDLOW** Motor Co. Ludlow. For CARS, TRACTORS and REPAIRS.

RM 0896. Cream. Reverse includes "Carfax. City of Oxford Motor Services Ltd. in association with G.W.R." and refers to Conditions of Issue.

   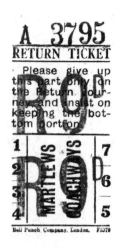

1553. Pale Green. Reverse blank.

C7286 Grey. Red overprint. Reverse: HAVE **TEA** AT **LEOPOLD'S**. The Popular Confectioners. Coliseum Cafe, Albion Street, 85 The Strand & 141 High Street, CHELTENHAM. Tel. 3710

(G8103) Buff. Reverse advises Modern Coaches are for hire etc. from "Poole's Coachways Ltd. High Street, Alsagers Bank, Stoke-on-Trent. Phone: Audley 245".

(A 3795) Light khaki with red overprint. Note different pattern of ticket. Reverse advises 'Luxury Coaches' available, MARTLEWS COACHWAYS 'Phone: Oakengates 207.

(Cr 7590) Pale Green. Reverse: Always use - GOODWIN's **EXTRA** SELF-RAISING **FLOUR**. It must be the Baby Picture Bag.

(8 Cy 6113) Beige, "1, Ordinary, 1d" in red. Reverse: *Sutton Park Throughout 'SUMMER TIME' on Saturday Afternoons Sundays & Holidays SERVICE No. 29 runs direct to BANNERS GATE entrance*

(Cb 9014) Apricot. Edmondson card. Reverse blank.

(H 6633) Sage with red overprint. Reverse: "Book your private party with Churchbridge Luxury Coaches Ltd. Phone: Cheslyn Hay 450".

(UA 1416) Grey-Green. Reverse: Buy **MORE** SAVINGS CERTIFICATES

(3864) Grey green. Note instruction at bottom. Reverse duplicates information on front plus "Phone . . . Hereford 2544".

93

A batch of less-than-usual tickets; but it is true that there was a ticket for every purpose so their variety is infinite.

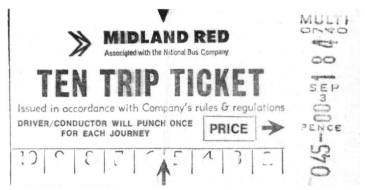

Ai 56794 Cream. Red type. Blue ink used in Setright machine. Reverse includes: "NOT VALID unless printed by machines at time of purchase. Defaced or mutilated tickets will NOT be accepted".

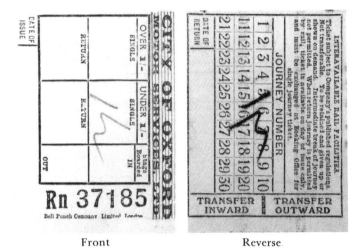

Rn 37185 Bellgraphic. Cream, red overlay. Entry made by hand.

(4402) Willebrew ticket. Mid-green. Reverse blank. Used for 'joint' services.

(2085) White. Type red and blue. The reverse of this rarity carries a photograph of the bus and the dates 1901 and 1953.

(FZ 2585) Pale green. Reverse: GAS DEPARTMENT To be used by Employees on OFFICIAL BUSINESS only.

(AE 6219) Mid Green. Reverse: *Let Worthington's feed you* LEICESTER'S LEADING GROCERS

Motor Bus Fittings.

Combined Folding and Sliding Door Opening Arrangement.

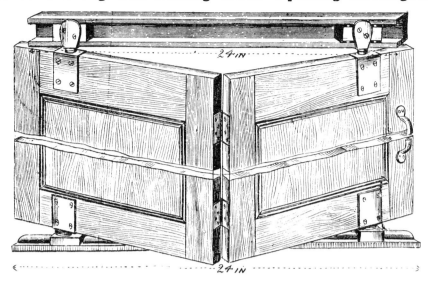

No. 5322. **£1 17 6** per set, including Steel Runner Bar, Brass Swivel Plates, Brass Bottom Runner, two Hinges and Handle.

Door Opening Arrangement for One-Man Bus.
REGISTERED DESIGN.

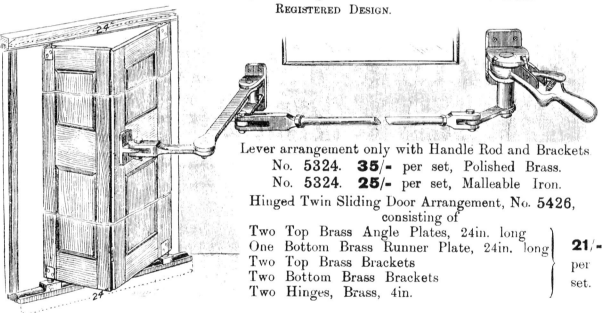

Lever arrangement only with Handle Rod and Brackets.
No. 5324. **35/-** per set, Polished Brass.
No. 5324. **25/-** per set, Malleable Iron.

Hinged Twin Sliding Door Arrangement, No. 5426, consisting of
Two Top Brass Angle Plates, 24in. long
One Bottom Brass Runner Plate, 24in. long
Two Top Brass Brackets
Two Bottom Brass Brackets
Two Hinges, Brass, 4in.
21/- per set.

Emergency Door Opening Device for Motor Omnibus.

No. 5323. **17/6** per set, Polished Brass Fittings, including Teak Bar and Lift.

JOHN BUCKINGHAM, BRUM'S BUS BUILDER

While Birmingham was known as 'The Toyshop of the World' due to a multiplicity of small items manufactured, nonetheless there were firms working in both the heavy industries and the assembly businesses. One coachbuilder was John Buckingham Ltd of Bradford Street, Birmingham. To judge by contemporary adverts they seem to have specialized in funeral hearses, but were also well known in the early 1920s for both lorry bodywork and complete bus bodies. As a company they never achieved the fame in p.s.v. circles of Eastern Coach Works; Park Royal or Massey, but where firms took a liking to their products they seem to have remained loyal until their designs became too outdated. A. Foley and W. Kendal of Preston, Lancashire, purchased their first new bus - a charabanc based on a Scout 40 h.p. chassis in 1915 but formed themselves into a Limited Company, the Pilot Motors, in 1920 transferring their by then five-strong fleet to the new company. Four of these had Maudslay chassis and from then until 1922 the Maudslay/Buckingham combination was favoured, a further eight vehicles following by the end of 1921. CK 3432 had a 32 seat bus body on Maudslay chassis No. 3592 type SB. (Strange that Bedford chose the same model designation for a chassis in 1950; one that followed the Maudslay philosophy). But all was not well with The Pilot, CW 3432 being sold out of the fleet in May 1926, and the firm was taken over by Ribble Motor Services Limited on 23 October 1926, by which time they operated a mixture of entirely Leyland bodied Leyland and Maudslay chassis.

The bottom photograph poses a mystery as its' serial number follows that of the Maudslay but the radiator profile is that of either the Associated Equipment Company (AEC) or the Daimler Motor Company. The chassis is quite remarkably straight and one would guess the body to be a 30-32 seater. The lack of any lettering is odd, the only known Buckingham demonstrator was in fact a char-a-banc in the classical multi-doored mould. Buckinghams were always favoured by the Great Western Railway for contracts, which says a lot for their workmanship, typically they bodied vast numbers of Burford chassis for the railway in 1924/5 and at least 75 Maudslay ML3 chassis in between 1927 and 1929. (Illustrated on page 97).

MIDLANDS INGENUITY

A design feature of 1920s and 1930s coaches was the flap, slide or lid supposed to cover the step-well. This scheme of 1937 may have been one answer but that "simple and light arrangements of levers" undoubtedly added to the draughts in the vehicle and with wear, made the cacophony inside the saloon rather worse than previously.

CONCEALING THE SLIDING-DOOR STEP WELL

AN OLD PROBLEM APPROACHED FROM A NEW ANGLE IN A LATERALLY SLIDING COVER FOR ATTACHMENT TO SLIDING DOORS.

The covering of the gap in the floor at the entrance of a coach has always been a problem that has vexed both the bodybuilder and the operator alike. Bodybuilders have for some time been alive to the necessity of a suitable cover and various methods have, from time to time, been devised. In most cases, however, they have had to be abandoned after a more or less lengthy trial.

The drawbacks have usually outweighed the advantages and, to-day, a step-gap cover is rarely seen. Another reason for this change is that the known devices were unsuitable for use in connection with the sliding door, which is now generally employed. So far as the writer is aware, no previous attempt has been made to conceal the well of a coach fitted with a sliding door.

For automatic actuation, it has generally been an accepted principle that a cover should be operated by the door movement, and there is no better or simpler method known. The earliest examples took the form of a ledge fixed on the inside of the door, at this period all doors being hinged. Variations of this arrangement were devised, but they applied solely to the hinged door and could not be employed with the sliding pattern.

The writer recently encountered this problem when building some coach bodies and cast about for some means for overcoming the difficulty. He evolved a scheme whereby the operation of opening and closing the door caused the cover to slide laterally between the cross-bearers and under the floor.

Operation is by a simple and light arrangement of levers. The scheme was worked out for both hinged and sliding doors. The movement of the cover is the same in both cases, but, naturally, the lever movement varies considerably. In each case only one end of the lever is attached to the door.

A laterally sliding cover to conceal the step well of a sliding door is now marketed by Messrs. Hallam, Sleigh and Cheston.

The complete fittings are manufactured by Messrs. Hallam, Sleigh and Cheston, Bagot Street, Birmingham. No special construction of the doors or underframing is necessary, and the cover can be fitted to an existing entrance. When the door is either open or closed the mechanism is out of sight, only the lever being visible during operation, as shown in the accompanying illustration. It is necessary that the gear should be protected from mud by light boarding fixed under the cross-bearers.

QUADRANT.

MIDLANDS MADE... or partially so!

When West Bromwich Corporation found their tramway services did not cater for the population around Greets Green and the rapidly growing area of All Saints' they decided to try buses as a prelude to trolley-bus services. Four Albions were purchased and three entered services on 27 July 1914, the fourth following in September. October 26 they were requisitioned for war work. Four electric (battery) vehicles were purchased and were so unreliable that by 1 March 1918 a doleful note in the local newspaper read "The local bus service suspended itself this week, by a breakdown". Four Tillings-Stevens TS3 chassis were purchased as replacements arriving in 1919. Frugally three of these were bodied using the components of the battery buses, but the fourth - No.5 shown here received a locally-built Roberts body seating 29 passengers. Incredibly this primitive machine lasted until August 1930.

The Birmingham-built bit of this primeval trolleybus, No. 240 in the Bradford City Tramways fleet, is completely invisible! This photograph was taken on the inaugural run, 20 June 1911 when the first fare paying trolleybus route in the UK was opened.

Two vehicles (240 and 241) were supplied by the Railless Electric Traction Company, having a 13' (3.96m) wheelbase and being 7' (2.13m) wide overall. Power for each came from two 20 h.p. Siemens motors and a Hurst Nelson 28 seat body covered the Alldays & Onions chassis. Alldays (correct name Alldays & Onions Pneumatic Engineering Co) built their first light commercial vehicles in 1906, and by 1911 had a 5-ton chassis available, upon which the trolleybus was assembled at their Matchless Works. The power method was interesting as the motors each drove a rear wheel via worm gearing to a countershaft and thence by chain.

A hand tram-type controller was fitted.

This rather unorthodox vehicle was owned by a firm called Land Liners Ltd of Edgware and was designed as one of a pair which in 1928 would carry 21 passengers in two and four berth cabins (ten sleeping below and eleven up top). All facilities were available including an attendant to provide both hot drinks and a full English breakfast from his galley - or kitchen, for it was equipped with the most modern fittings. The toilet (with mahogany seat) and flushing mechanism together with a handbasin, mirror and vanity unit was downstairs. The fares (15/6d single and 30/- return) were far too low for bed and breakfast and although Strachans & Brown built excellent bodies, they were heavy and possibly too much so for the rather feeble 35 h.p. six-cylinder sleeve valve engines fitted to their Guy FCX three axle chassis. Land Liners appear to have faded away in 1929.

CUL 319 is, on the face of it a perfectly orthodox London Trolleybus. Seen here in the 1950s towards the Barnet terminus she was one of a class of 100 with AEC chassis. Her unorthodoxy (if that is what it was) lay in that the bodies of the C3 class were built by the Birmingham Railway Carriage & Wagon Company of Smethwick. The BRCW was first established in 1878 and were particularly well-known in the export field, although they did build a few tramcars. Typically between 1925 and 1932 they supplied the Wagon-Lits Company with 88 sleeping cars, 31 dining cars and 47 Pullmans, all exported. They also manufactured diesel locomotives (B.R. classes 26, 27 and 33) and DMU vehicles. In 1963, for various reasons, they ceased to trade. The full run of route 645 was from Barnet, via Cricklewood and Edgware to Canons Park.

Still on a London Transport theme but far more unusual was this class of 84 vehicles known as the GS. The radiator cowling proudly carries the 'Feathers in our Cap' symbol of Guy Motors, but how much of them was of Guy origin is rather dubious. The chassis is, as can be seen (remarkably for 1953) of normal control pattern, and uses mainly Vixen components. Classified as NLLVP by Guy Motors they have a wheelbase of 14'9" (4.5m) and are powered by a Perkins P6, 6 cylinder diesel derated to 65 h.p. coupled to a 4-speed crash gearbox. The cowling is by Briggs Motor Bodies and the bodywork, seating 26, was built by ECW of Lowestoft. At least 17 are preserved of whom two are seen here at Dartmouth Park, 13 May 1990.

An oddity in the London Transport fleet was G436, purchased to provide a direct comparison between the 'home-grown' AEC/Park Royal RT class vehicles and one of the most powerful of the provincials. Seen at Peckham Rye G436 had a Meadows 10.35 litre engine coupled to an air-operated pre-selective gearbox. The body was built by Guy at Wolverhampton although the design was probably licensed from Park Royal. Entering service in January 1950 she was allocated to Old Kent Road garage to run on the 173 circular. Long outlasting her 'utility'-bodied sisters, KGK 981 ran until February 1955 and had the melancholy distinction of being the last double-deck Guy to be used in London.

A beautifully unhurried scene at Bovey Station in the summer of 1926. The Great Western Railway (whatever their other failings) were quick to seize on the chance of increased revenue from the operation of tourist coaches. Special trains were quite commonplace, even into the 1950s and, certainly from London, "Mystery Tours" often incorporated a coach trip from one railhead to another or, as here at Bovey, to a local sightseeing venue.

T7692 for Haytor Rocks was a 45 h.p. AEC with GWR (Swindon) built body, registered 1.1.20, but XY 2110 for Becky Falls and Manaton was a Burford with an 18 seat Buckingham of Birmingham charabanc body first registered on 24 April 1925. At the rear L6307 is another AEC/GWR combination. The tour to Haytor operated between January 1924 and December 1928, Becky Falls September 1925 - December 1928.

One of the more unsuccessful Midlands vehicles was the Sunbeam Pathan, first introduced in 1929. Designated the SF4 only 20 chassis were ever built between then and 1934, despite a fairly advanced design and specification. It may be that petrol engines were falling into disfavour although the Sunbeam's 6597cc (nominally 37.2 h.p.) had rather puny horses with the vehicles being sluggish on rising gradients.

The first example, 190, one of a batch of three with 31-seat Taylor of Norwich bodies, was delivered to Wolverhampton Corporation Transport in December 1930 and is seen when new. Her sister, 189, is scurrying along the Wolverhampton-Bridgnorth road (A454) in 1931. Both were relatively shortlived, and after some spells of idleness were de-licensed on 30 November 1938.

SANKEY
Motor Coach Bodies

of the latest types, to suit any make of chassis. From 10 to 35 seaters to customers' choice.

IMMEDIATE DELIVERIES

can be given in the shell finish at prices ranging from £80 to £170

Any type of body can be supplied. Send your enquiries to the Pioneers of the MOTOR COACH BODY.

JOSEPH SANKEY & SONS Ltd
(Makers of the All-Steel Wheel, One-piece Domed Wings, Panel Stampings, etc.)

HADLEY CASTLE WORKS, WELLINGTON
SHROPSHIRE

WALSALL DOUBLE-DECKERS
A new look in 1951

128 is a Leyland Titan PD2/1 chassis with a Park Royal body in fairly original condition, delivered 1951.

231 a Guy Arab III 6LW, has a body which, as delivered in 1952, was almost identical with 128, but which was modernized in the 1960s. Behind 231 is 884 with a full-height Willowbrook body, on the lowest avilable 1960 commercial chassis, the Dennis Loline II.

Walsall Corporation commenced operations in 1904 as a tramway operator, moved into buses on 23 May 1915 with a route to Bloxwich and Hednesford using petrol-electric vehicles. After abandoning some tramways and replacing them with motor vehicles, from 1931 trolley-buses were very highly regarded with, rather improbably, the last route extension being completed in 1963. The fitters at Bloxwich Road Garage seemed to have had a fairly free hand and by comparison with the uniformity of London buses, Walsall's verged on the downright eccentric.

The new overall length of 27 ft. for two-axle double-deck motor buses has given bodybuilders just that little extra to play with in their designs, and one such example of this new freedom is shown in a contract for the Home Market on which Park Royal Vehicles are currently engaged, that for the Transport Department of Walsall Corporation. For this operator Park Royal are building a fleet of fifty double-deck buses 7 ft. 6 ins. wide and of the newly permitted maximum overall length of 27 ft. 0 ins., half of the total number of chassis upon which the bodies are mounted being supplied by Guy Motors and the remainder by Leyland. The body construction is of the composite type, and with the design featuring a full-width driver's cab the vehicles present a most pleasing appearance.

The construction features an underframe entirely of English oak, with all the cross bearers centre flitched with 3/16 in. steel plates. The remainder of the framework consists of selected pitch pine or ash suitably reinforced and braced with flitch plates, gussets and brackets, etc. Following Park Royal standard practice all materials used receive a thorough protective treatment to ensure the long life of the bodywork.

The full-width driver's cab is semi-floating and built integrally with the body, and the framing, which is built over the radiator, has an attractively styled frontal grille embellished with polished alloy mouldings. To facilitate engine maintenance this grille is made completely detachable. A full-depth sliding cab door is fitted either side of the body, mounted on self-aligning runners, and with sliding windows incorporated for signalling and ventilation purposes. The offside windscreen is a two-panel type, with the top half opening outwards on robust quadrants, whilst the nearside is fitted with a fixed screen. Both screens are framed in brass sections, chromium plated. Being of exceptional depth these screens afford both driver and passengers an excellent range of vision.

Exterior panelling is of aluminium sheet with all panels screwed to the framing and secured at the joints with cover strips and mouldings. Four stainless steel half-drop windows are fitted in each saloon. Other saloon windows are glazed direct into the body framework with rubber glazing channel. Continuous aluminium louvres are fitted over all saloon side windows and also over the upper saloon front end windows. Four "Roe-Vac" extractor type ventilators are incorporated in the upper saloon roof complete with chromium plated interior grilles. An arrangement for permanent ventilation is provided in both saloons. Provision is also made for ventilation in the driver's cab.

Seating accommodation is provided for 56 passengers, thirty in the upper saloon and twenty-six in the lower. Seats are of the steel tubular "service" pattern with chromium plated top rails and incorporate corner grabs. Cushions and squabs are Dunlopillo filled and trimmed in brown Connolly hide with fully pleated cushions. Brown leather-cloth to match finishes the seat backs. The driver's seat is a semi-bucket type upholstered in brown hide and mounted on fully adjustable seat fittings. The interior finish of the saloons is in a two-colour scheme of brown and white with all finishers walnut polished.

Saloon floors are covered with 4.5 mm. linoleum of light brown colour extending to the seat rails on radiused metal coves designed to facilitate cleaning. Cowls and traps are fitted where required for access to chassis components. The gangway and the floor between the seats are provided with hardwood slats painted a dark brown.

The saloon ceilings are finished in white enamel, and in both saloons and at the platform polished stainless steel floor-to-roof stanchions and handrails are fitted. The lower and upper saloon bulkheads are finished with 3 mm. brown linoleum, with the sill at the upper saloon front bulkhead edged with fluted aluminium.

Interior illumination is provided by open type reflectors accommodating 12w. pearl bulbs, ten in the lower saloon, fourteen in the upper, and one over the rear platform. All saloon lamps are mounted in standard lighting cove panels and controlled by a switch and fuse box in the driver's cab. A rear red light is incorporated in the vent panel on either side over the rear windows. A signal bell positioned in the cab is operated by a cord fitted to the lower saloon roof and extending to the upper saloon for operation from beside the top-deck stairwell fender rail. A flush type trafficator is fitted either side of the body at the front with a two-way switch control in the cab.

An illuminated destination indicator surmounted by a route number box is arranged centrally at the front of the body and both are designed for operation from the upper saloon.

Finished in Dulux synthetic enamels, the exterior is painted in a colour scheme which enhances the appearance of these extremely well-proportioned vehicles.

DUNLOP LATEX PRODUCTION
Further Supplies from Ceylon

A new rubber factory is being opened in Ceylon by the Latex Corporation of Ceylon Ltd., a subsidiary of the Dunlop Rubber Company, recently registered with an authorised capital of 4,000,000 rupees, equivalent to £300,000. The factory is at Kalutara, a rich rubber-planting district about 27 miles from Colombo.

Rubber estates in the surrounding areas have been asked to offer their field latex which the Corporation will buy from them and from other rubber estates in Ceylon under contracts over various periods at a monthly average price, less two or three cents for transport to factories and the prevailing cess. Suppliers are also promised fifty per cent. of the net profits from the venture, divided according to the quantity of latex supplied.

The processing of field latex into Dunlop concentrated latex is the sole activity of the new Corporation. On behalf of the parent company, Dunlop Plantations Ltd. are supervising these activities of the factory and also the sale of the concentrated latex to the rubber manufacturing industry for the various products, mainly Dunlopilo, made from it.

DOO DAIMLERS FOR COVENTRY CORPORATION
MAR - MAY 1968 (MONOBUS)

The first of the 18 new Daimler CRG6LX Fleetline buses, with the specially-designed ECW double-deck bodies seating 72 passengers, started arriving in the New Year. They had front entrances and centre staircases, vestibules and exits and were fitted with automatic ticket cancellors to be used by passengers presenting pre-paid tickets. They were the first double-deckers in the country built specifically for one-man operation and, as they were ahead of their time, needed special MOT dispensation before use as such. They were finished in ivory livery with Arabian red bands. Equipped with front (now three-digit) route and destination boxes "in-line", a new feature was a small nearside route number box above window-level by the front door. Fleet numbers were 23-40 (KWK 23-40F). They were introduced into service for crew and passenger acclimatization and "proving" on services proposed for one-man operation.

A rather unorthodox purchase for a Corporation fleet were three Daimler Freeline D650HS vehicles purchased by Coventry in 1959. Numbered 401-3 (XRW 401-3) they were fitted with Gardner 6 HLW engines, which were at least recognisable to Coventry's fitters, and reasonably luxurious Willowbrook 41 seat coach bodies. Quite how much work they did in this guise is uncertain but by 1965 they were relegated to bus service on route 19, after being fitted with wider doors, destination boxes and a luggage pen, this latter reducing capacity to 39 seats. In 1967, as part of a conversion for one-person-operation (O.P.O.) air operated doors were fitted. Withdrawn prior to the formation of the West Midlands P.T.E. their lives had been, to say the least, unusual. No. 402 is seen on an excursion when still virtually new.

Seen sometime between 1916 and 1919, HD 201 was one of six 16-seat Daimlers acquired by Yorkshire (Woollen District) Electric Tramways in 1913-14 to provide services from Scholes to Cleckheaton and between their Ravensthorpe tram terminus and that of Huddersfield Corporation Tramways at Bradley, where the physical difference in gauges (which so bedevilled tramways interworking) contributed to an unwillingness to expend the money on trackwork.

The two last vehicles (HD 200/1) had the luxury of enclosed cabs and for that reason were converted into maintenance wagons in 1921, whereas their four earlier sisters were disposed of. As tramway overhead vehicles they were still extant in 1929.

Photograph taken at Bradley Terminus with the tramway shelter to the left.

The Daimler COG series vehicles introduced in 1933 were the first from this manufacturer to incorporate diesel (oil) engines, hence C = Commercial, O = Oil, G = Gardner, the engine manufacturer; 1900 (including an AEC engined variant, the COA) were built by 1940 when production ceased. In 1943 permission was given to Daimler to produce a heavier version of the COG chassis, using iron to replace scarce alloys. Logically this became the CWG (W = Wartime) series when fitted with Gardner engines; 138 were built, together with 125 Daimler (CWD) engined variations, and 1116 with AEC engines (CWA). After the war, in 1946, as materials became available, a CV design appeared, with the V standing for Victory. Initially four engines were offered, Gardner 5 and 6 cylinder (CVG 5 & 6), the Daimler CD6 (CVD6) and the AEC 7.7 litre 6-cylinder design (CVA 6). Variations on the theme remained in production until 1965.

In September 1948 double-deck vehicles were provided to carry passengers to the British Exhibition, held in Copenhagen, as part of the post-war celebrations. Each vehicle as can be seen was fitted with a special Daimler radiator scroll, and in the case of the leading Glasgow vehicle 157 a CWD 6 of 1945, her Brush bodywork was specially painted, including coats-of-arms on the upper deck panels and a proud "City of Glasgow" above the lower deck windows. "GB" plates and the semaphore signals (near the driver's cab) were retained for some time after. Not all the other vehicles in the photograph even have chromed radiators, but all are 1948 built, CVD6 chassis carrying Brush bodywork, except the second (Birmingham) bus which is bodied by MCW. The others are from Bradford, Stalybridge, Nottingham and Salford. Poignantly, the last Leylands to bear a famous type name - the 'Lion' - were built in Denmark by their associated company DAB albeit with a production run of less than a year, finishing just 40 years after our Daimlers (later Leyland owned of course) went to Copenhagen.

APPENDIX 1   MCW drawings: 1952 LT RF Class

SPECIFICATION
L.T.E. "Central" Bus
M.C.W. Single-deck Bus on A.E.C. Regal Mk.IV Chassis

DIMENSIONS

Overall dimensions are 30' 0" long by 7' 6" over wings by 10' 5" unladen height.

CONSTRUCTION

The body is of Metro-Cammell patented metal construction embodying patented tubular steel pillars and inner riveted structure panels of high tensile aluminium alloy.

Outer panels consist of aluminium, "Pop" riveted to horizontal and structure members.

The body is permanently bolted to the chassis at the front cross member and riveted to eight main outriggers provided on the chassis.

FLOOR

The floor is constructed of ¾" tongued and grooved boards and the gangway consists of hardwood slats with cork tiles under the seats. The coving panels are covered with linoleum.

ROOF

The roof is of the double-skin type with interior panels of 3/16" plywood extending transversely in bay lengths from cantrail to cantrail in one piece. The exterior panels are of 18 G. aluminium.

DRIVER'S CAB

The windscreen in the driver's cab is of the recessed type to minimise glare.

The driver's seat is fully adjustable, both vertically and horizontally.

A sliding window is provided at the side of the driver for signalling purposes.

A heater is fitted in the driver's compartment.

A partition at rear of driver and a waist-high partition at nearside of driver incorporating a hinged door is also provided.

ENTRANCE & EXIT

The entrance and exit which is forward of the front wheels is without doors.

WINDOWS

Eight half-drop windows of the high-level winding type are fitted - four on each side of the vehicle.

The window pans are of steel. All windows are glazed in Triplex toughened safety glass except the emergency escape window which is of plate glass.

Glazing is of the simplastic rubber type.

VENTILATION

One intake ventilator in the centre of the roof and two extractor ventilators in the rear dome are provided in addition to the above half-drop windows for ventilation purposes.

DESTINATION INDICATORS

Route indicator boxes are fitted in front and rear canopies with mechanical gearing.

SEATING

Seats are provided for 41 passengers. The seat frames are of the tubular steel type with cushions of Dunlopillo. All seats are upholstered in moquette to the pattern of London Transport Executive.

HANDRAILS ETC.

For standing passengers ceiling handrails are provided between the longitudinal seats over the front wheelarches. Entrance handrails are covered white Doverite.

INTERIOR FINISH

Inside lining panels from seat rail to waist are covered with leathercloth.

Covered shrouds are fitted around the windows. Interior roof panels are enamelled white.

INTERIOR LIGHTING

Interior illumination is provided by 25 lamps in quickly detachable fittings of L.T.E. patented design.

MISCELLANEOUS

Provision is made for the installation of trafficators.

The rear registration number plate incorporates direction arrows, and a stop lamp is also fitted.

The filler cap, which is positioned immediately below the windscreen, is accessible from the outside and is covered by a motif flap of L.T.E. design.

A bell cord extends the full length of the saloon.

The exterior is finished to the requirements of the London Transport Executive.

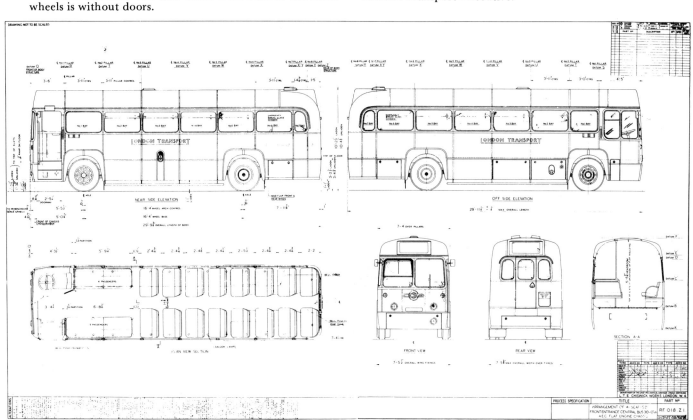

L.T.E. "CENTRAL" BUS.

APPENDIX 2

Pride o' Wolverhampton: Sunbeam Trolleybuses — Andy Simpson

'THE SAFEST FOUNDATION FOR A TROLLEYBUS IS A SUNBEAM CHASSIS' - Sunbeam Brochure, 1954.

The web of trolleybus wires covering large parts of the industrial Midlands from the 1920s to the 1960s was to a large extent served by locally built vehicles from the Sunbeam stable, which latterly included Guy and Karrier vehicles also, all produced in Wolverhampton.

An old established company with its origins in bicycle manufacture Sunbeam produced trolleybuses at its Moorfield Works, Wolverhampton, from 1931 onwards, the 1931 prototype the MS1 six-wheeler, going to Wolverhampton Corporation with the first production vehicles in 1933.

Sunbeam became part of the Rootes group in July 1935; Rootes had previously purchased the commercial vehicle builders Karrier Motors Ltd of Huddersfield in August 1934, and transferred Karrier trolleybus production to Sunbeams' Wolverhampton works, with Sunbeam and Karrier designed trolleybuses being built side by side at the same works. Rootes sold Sunbeam in 1946 and it became the Sunbeam Trolleybus Company Ltd in 1948, being sold to Guy Motors Ltd that same year, the Guy trolleybus business being handed over to Sunbeam. The last Karrier chassis appeared the same year. The last Guy chassis delivered in April 1950, was, appropriately enough, delivered to Wolverhampton as their last new trolleybus, the still extant No. 654, currently stored in Northants.

All future production was marketed as Sunbeam, manufacture being at Guy's Fallings Park, Wolverhampton Works from 1953 until production ended, after a four year gap, with a batch of six vehicles for Coimbra, Portugal in 1966. Despite a lack of orders, the range remained available until mid-1967 at least. By then Sunbeam, and its parent company Guy, were part of the Jaguar empire, having been taken over in October 1961 when Guy entered a financial crisis fuelled by development costs of the air-suspension Wulfrunian and unwise involvement in the South African market.

The Guy/Karrier/Sunbeam chassis range during the period 1931-1966 was nothing if not varied:

GUY

BT (2 axle) 1928-1938 and 1949-1950 (the latter being 50 for Wolverhampton)

BTX (3 axle) 1926-1939 and 1947-1949 (the latter being 70 for Belfast)

KARRIER

E4 (2 axle) 1936-1942. 57 built.

E6/6a (3 axle) 1936-1940. 174 built.

Until 1940, these Karrier models were built more or less separately from the Sunbeam models at the same Moorfield works, with little standardisation of equipment. Huddersfield 541-68, CUH 741-8, DUH 49-68 - Sunbeam MS2s of 1947/8 - have Karrier nameplates - the last time the Karrier name appeared on full-sized passenger vehicles.

KARRIER/SUNBEAM 2-AXLE

F4 1948-1953 26', 27' or 30' chassis. Over 250 built - single or double-deck bodywork.

F4a 1953-1961. 30' chassis. 65 built. Incorporated many Guy Arab IV motorbus components. Variety of wheelbases available. 63 built for U.K. service, plus 2 show chassis, later dismantled.

MF1/2/3 1934-1942. 120 built.

MF2B 1952-1966. 27' chassis - 30' from 1954. 55 built for U.K. service 1953-1962. Final batch of six to Coimbra, Portugal, 1966. Originally introduced as a single deck export model, as was the MF2R version.

In the 1950s these chassis offered a choice of electrical equipment from BTH, Crompton-Parkinson, General Electric or Metro-Vick.

SUNBEAM 3-AXLE

MS1/2/3 1931-43 and 1947-1951. 27' or 30' chassis. Over 500 built.

S7 (8' wide)/S7A(7'6" wide)/S7B(Foreign Service) 1948-1960. 23 built for U.K. service to 1959, plus 60 for foreign service, the last to Johannesburg in 1960.

KARRIER/SUNBEAM

W4 1943-1949. 2 axle wartime 'utility' design for Ministry of Supply Specifications "suitable for all", using an angular timber-framed body. Sunbeam/Karrier was the sole U.K. trolleybus manufacturer in the latter part of World War 2. Over 200 of these 'utilities' were built, being rather more 'Sunbeam' than Karrier in production.

Unless otherwise stated, the above figures and dates include production for both U.K. and foreign users.

There was an initial intention to build a trolleybus version of the air suspension Guy Wulfrunian, introduced 1959, but this scheme was not proceeded with.

© 1992 Andy Simpson

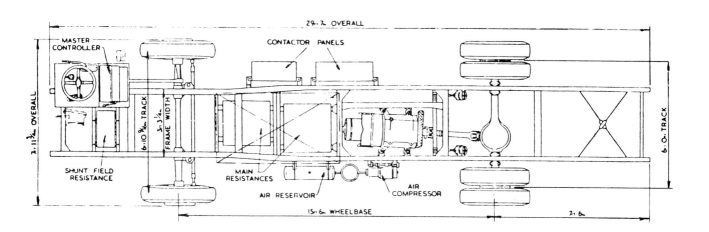

INDEX TO CHASSIS & BODY MANUFACTURERS

ADC 59
AEC Frontis, 2, 3, 6, 10, 18, 20, 49, 63, 66, 81, 98, 99, 104
AILSA 67, 68, 69
ALBION 20
ALEXANDER 53, 54, 67
ALLDAYS & ONIONS 98

BARTON (BTS) 47
BECCOLS 3
BEDFORD 20, 66, 76
BELLHOUSE HARTWELL 81
BERLIET 18
B.R.C.W. 98
BRISTOL 67, 68, 69, 76, 80
BRUSH Frontis, 37, 38, 40, 66, 75, 82, 83, 103
B.T.H. 40
BUCKINGHAM 48, 96, 97, 99
BURFORD 97, 99
BURLINGHAM 17, 20

CHEVROLET 42
CLARK 81
CLARKSON 4
COMMER 10
CROSSLEY 41, 65

DAIMLER Frontis, 20, 24, 26, 48, 54, 63, 66, 67, 68, 69, 75, 82, 85, 102, 103
DAVIES 85
DENNIS 35, 43, 44, 66, 81, 100
DODSON 12
DUPLE 2, 3, 10, 18, 20, 35, 53, 63, 66, 81

E.C.W. 30, 54, 98, 102
ENGLISH ELECTRIC 53

FODEN 10
FORD 10, 20

GABRIEL 95
GARFORD 5
GILFORD 8, 9, 44
GLOSTER-GARDNER 2
GURNEY NUTTING 81
GUY 26, 27, 37, 39, 49, 62, 82, 83, 85, 98, 99, 100, 101, 105
G.W.R.(SWINDON) 99

HALL, LEWIS 6
HARKNESS 26
HURST, NELSON 98

KARRIER 36, 38, 105

LEYLAND Frontis, 8, 9, 11, 17, 20, 27, 35, 38, 43, 47, 54, 55, 65, 66, 80, 81, 100
LGOC 6

MARSHALL 26, 48, 65
MASSEY 85
MAUDSLAY 2, 14, 20, 83, 96
MCW Frontis, 25, 37, 38, 48, 49, 54, 67, 85, 103, 104
MERCEDES-BENZ 55
METALCRAFT 10
METAL SECTIONS (OLDBURY) 27
MIDLAND RED 13, 35, 50, 51, 52, 53, 62
MINERVA 5
MORRIS-COMMERCIAL 42, 48, 49, 70
MULLINER 27

NCME 80, 85

PANHARD 2
PARK ROYAL 20, 38, 39, 40, 66, 67, 80, 85, 99, 100, 101
PLAXTON 10, 20, 34
PMT 55

RAILLESS ELECTRIC TRACTION 98
REO 34, 42, 43
ROBERTS 98
ROE 36, 37, 39
ROOTES 55

SANKEY 99
SENTINEL 84
SHORT 38, 49, 53
STRACHANS 49
STRACHANS & BROWN 98
STRAKER 4
STUDEBAKER 11
SUNBEAM 26, 37, 39, 40, 62, 99, 105

TAYLOR 99
THORNYCROFT 42, 81
TILLING-STEVENS 11, 30, 37, 98
TRANSUNITED 41

UNITED 9
U.T.A. 27

VAN-HOOL-McARDLE 67

WEYMANN 37, 38
WHITSON 2
WILKS & MEADE 66
WILLOWBROOK Frontis, 35, 37, 38, 85, 100, 102
WYCOMBE 8, 9, 44

YEATES 66, 81

> NOTE: To keep this index within manageable limits no attempt has been made to list entries in "Everyone A Little Gem".

Wolverhampton Corporation

Leicester City

M I

A ga
Some of the council

Northampton Corporation

Coventry Corporation